Long-Term Performance Prediction for PVC Pipes

Awwa
Research
Foundation

About the Awwa Research Foundation

The Awwa Research Foundation (AwwaRF) is a member-supported, international, nonprofit organization that sponsors research to enable water utilities, public health agencies, and other professionals to provide safe and affordable drinking water to consumers.

The Foundation's mission is to advance the science of water to improve the quality of life. To achieve this mission, the Foundation sponsors studies on all aspects of drinking water, including supply and resources, treatment, monitoring and analysis, distribution, management, and health effects. Funding for research is provided primarily by subscription payments from approximately 1,000 utilities, consulting firms, and manufacturers in North America and abroad. Additional funding comes from collaborative partnerships with other national and international organizations, allowing for resources to be leveraged, expertise to be shared, and broad-based knowledge to be developed and disseminated. Government funding serves as a third source of research dollars.

From its headquarters in Denver, Colorado, the Foundation's staff directs and supports the efforts of more than 800 volunteers who serve on the board of trustees and various committees. These volunteers represent many facets of the water industry, and contribute their expertise to select and monitor research studies that benefit the entire drinking water community.

The results of research are disseminated through a number of channels, including reports, the Web site, conferences, and periodicals.

For subscribers, the Foundation serves as a cooperative program in which water suppliers unite to pool their resources. By applying Foundation research findings, these water suppliers can save substantial costs and stay on the leading edge of drinking water science and technology. Since its inception, AwwaRF has supplied the water community with more than $300 million in applied research.

More information about the Foundation and how to become a subscriber is available on the Web at **www.awwarf.org.**

Long-Term Performance Prediction for PVC Pipes

Prepared by:
Stewart Burn, **Paul Davis**, **Tara Schiller**, **Bill Tiganis**, **Grace Tjandraatmadja**, **Mark Cardy**, **Scott Gould**, and **Paul Sadler**
Commonwealth Scientific & Industrial Research Organization (CSIRO), Australia
and
Alan J. Whittle
Iplex Pipelines Australia Pty Limited
35 Alfred Road, Chipping Norton, NSW 2170, Australia

Sponsored by:
Awwa Research Foundation
6666 West Quincy Avenue, Denver, CO 80235-3098
and
Commonwealth Scientific & Industrial Research Organization (CSIRO)
Limestone Avenue, Canberra, Australia

Published by:

Awwa Research Foundation
Advancing the Science of Water®

DISCLAIMER

This study was jointly funded by the Awwa Research Foundation (AwwaRF) and the Commonwealth Scientific & Industrial Research Organization (CSIRO). AwwaRF and CSIRO assume no responsibility for the content of the research study reported in this publication or for the opinions or statements of fact expressed in this report. The mention of trade names for commercial products does not represent or imply the approval or endorsement of AwwaRF or CSIRO. This report is presented solely for informational purposes.

CONTENTS

TABLES

FIGURES

FOREWORD

The Awwa Research Foundation is a non-profit corporation that is dedicated to the implementation of a research effort to help utilities respond to regulatory requirements and traditional high-priority concerns of the industry. The research agenda is developed through a process of consultation with subscribers and drinking water working professionals. Under the umbrella of a Strategic Research Plan, the Research Advisory Council prioritizes the suggested projects based upon current and future needs, applicability, and past work; the recommendations are forwarded to the Board of Trustees for final selection. The foundation also sponsors research projects through the unsolicited proposal process; the Collaborative Research, Research Applications, and Tailored Collaboration programs; and various joint research efforts with organizations such as the U.S. Environmental Protection Agency, the U.S. Bureau of Reclamation, and the Association of California Water Agencies.

This publication is a result of one of these sponsored studies, and it is hoped that its findings will be applied in communities throughout the world. The following report serves not only as a means of communicating the results of the water industry's centralized research program but also as a tool to enlist the further support of the nonmember utilities and individuals.

Projects are managed closely from their inception to the final report by the foundation's staff and a large cadre of volunteers who willingly contribute their time and expertise. The foundation serves a planning and management function and awards contracts to other institutions such as water utilities, universities, and engineering firms. The funding for this research effort comes primarily from the Subscription Program, through which water utilities subscribe to the research program and make an annual payment proportionate to the volume of water they deliver and consultants and manufacturers subscribe based on their annual billings. The program offers a cost-effective and fair method for funding research in the public interest.

A broad spectrum of water supply issues is addressed by the foundation's research agenda: resources, treatment and operations, distribution and storage, water quality and analysis, toxicology, economics, and management. The ultimate purpose of the coordinated effort is to assist water suppliers to provide the highest possible quality of water economically and reliably. The true benefits are realized when the results are implemented at the utility level. The foundation's trustees are pleased to offer this publication as a contribution toward that end.

Walter J. Bishop
Chair, Board of Trustees
Awwa Research Foundation

Robert C. Renner, P.E.
Executive Director
Awwa Research Foundation

ACKNOWLEDGMENTS

The authors of this report are are indebted to the following water utilities and companies for their cooperation and participation in this project:

Roy Brander, City of Calgary, Alberta, Canada
Russell Holmes, City of Ottawa, Ontario, Canada
Lucia Cade, Evan & Peck, formerly City West Water, Victoria, Australia
Dan Klopfer, Des Moines Water Works
Michael Wallis, East Bay Municipal Utility District, Oakland, California, USA
Doug Seargeant, Epcor Water Services, Edmonton, Alberta, Canada
Balvant Rajani, National Research Council Canada (NRC), Ottawa, Ontario, Canada
Solomon Tesfamariam, National Research Council Canada (NRC), Ottawa, Ontario, Canada
Greg Moore, SA Water, South Australia, Australia
Terry Anderson, South East Water, Victoria, Australia
Mark Heathcote, Earthtech, formerly Sydney Water, New South Wales, Australia
Jim Taylor, Tyco Water Plastic Pipeline Systems, New South Wales, Australia
Bob Walker, Uni-Bell PVC Pipe Association, Texas, USA
Peter Chapman, Vinidex Pty Ltd, New South Wales, Australia
Rod McCoy, Yarra Valley Water, Victoria, Australia

The authors would like to thank the following AwwaRf Staff and Project PAC members for their support and constructive contribution to the project

Stephanie Passarelli, AwwaRF
Jian Zhang, AwwaRF
Jose Rios, East Bay Municipal District
Stephen McKelvie, Parsons Brinckerhoff Quade & Doyle and
Michael Farrimond, UKWIR

In addition the authors would like to thank United Kingdom Water Industry Research for kindly providing a database of pipe performance characteristics for the United Kingdom Industry.

EXECUTIVE SUMMARY

Since the late 1970s, polyvinyl chloride pipes have been used in a significant number of new water distribution systems in North America. While premature fracture failures have been reported for (unplasticized) PVC-U pipes worldwide, with good design and installation practice, current performance standards indicate that water utilities should expect a minimum 50-year service life from a PVC pipe, with expert opinion suggesting that 100 years is a conservative estimate for a "properly designed and installed pipe".

Quantifying the in-service life expectancy for PVC pipes will enable greater confidence in their use and, as a natural outcome, will also influence long-term asset management and replacement planning strategies applied by water utilities.

The Awwa Research Foundation and CSIRO funded this study in order to be able to outline an approach to estimating in-service lifetimes for (unplasticised) PVC-U (commonly called PVC), (modified) PVC-M, and (oriented) PVC-O pipes, as well as the influence of manufacturing and installation practices on these lifetimes.

Pipe failure can occur by either causing water quality issues or by physical failure. These can occur due to chemical breakdown of the polymer structure, by structural failure of the pipeline, by fatigue, or by a crack growth mechanism leading to physical failure. The aim of this project was to asses the likelihood of chemical, fatigue, and physical failures of PVC pipes, and recommend processes to either estimate lifetimes under these processes or the steps that need to be taken to minimize the effects on lifetimes. As the factors determining structural failure (such as buckling) are well documented and generally rely on the changes of modulus with time, they were not included in this study.

For chemical failure, the requirements for water quality, ultra-violet (UV), and heat stabilizers were reviewed, and the necessary standards tests to ensure adequate lifetimes were researched and recommendations made. In the case of fatigue, most water distribution pipelines are not subjected to fatigue, except for the diurnal changes in pressure, which are usually benign. Fatigue is a special design case and, where it occurs, specific design criteria are required. Significant research has been carried out on the performance of PVC under fatigue situations and this study did not attempt to duplicate that research. However, the current research has been reviewed and design guides have been recommended for the North American market. For physical failures, the current performance requirements for manufacture have been reviewed and recommendations made. In addition, a fracture mechanics-based model to predict the conditions under which pipe failure will occur in service has been developed, and this has been calibrated against failure rates recorded in a number of North American and Australian utilities. This approach requires the fracture properties that govern failure in service to be quantified along with the expected loading conditions for a buried PVC pipe.

CONCLUSIONS AND RECOMMENDATIONS

Water Quality

Evidence to date indicates that metal stabilizers migrate into water from manufactured PVC pipes. Research on the toxicity of the metals used in the heat stabilizers has resulted in the setting of a maximum contaminant level of 0.010 mg/L (ppm) for lead in Australia and Europe, and a target level of 0 mg/L (ppm) in the USA (0.015 mg/L or ppm action level). Levels for organotin have not been established by the World Health Organization (WHO) and national regulatory bodies due to insufficient evidence confirming its toxicity. Calcium/zinc (Ca/Zn) stabilizer systems are considered non-toxic and safe for use. For tin- and organic-based stabilizers, further investigation into their toxicity may be required, however, if a maximum contaminant level is applicable, current information does not suggest that these substances are harmful to humans in the concentrations detected in potable water.

To ensure water quality is maintained, pipes manufactured for use with potable water are required to pass ANSI 61, which is very thorough in its requirement for health hazard evaluation. However, the extraction methodology could be simplified without compromising hazard determination. Regarding the use of solvent cement joints, it is recommended that evaluation of the migration of stabilizers be determined for small diameter pipe, and that consideration be given to the time required for the joint to cure before putting the pipe into service, as solvent cement appears to increase stabilizer mobility.

When pipes are installed and commissioned, flushing the pipe under turbulent flow before use is strongly recommended to remove excess depleted stabilizer from the pipe surface. Moreover, swabbing is standard good construction practice as it removes unhygienic materials that may have been deposited in the pipe during installation (Holmes, 2004). Maintaining water quality should not be an issue for PVC pipes that pass ANSI 61 and are swabbed prior to being placed in service.

Heat and Weathering Resistance

The requirements for heat and weathering resistance of PVC pipe are extremely well documented, and any procedure for predicting weathering and thermal degradation effects on performance should be based on the key properties that govern pipe service lifetime. They should include (for example) impact strength, fracture toughness, and dehydrochlorination tendency as measured by pH. However, further development of the models to predict long-term properties from short-term tests, or accelerated exposures is necessary. It is important to note that these predictions will be formulation specific, and cannot be applied to all PVC materials. Numerous citations in the literature clearly demonstrate that a clear correlation between natural weathering/heat aging of PVC to artificial methods has not been established and if accelerated methods are to be used, this correlation needs to be determined if assessment methods based on artificial exposures, or property extrapolation are to provide accurate indications of PVC pipe performance.

While the failure mechanisms of pipes exposed to heat or UV exposure is relevant to long-term performance under aggressive installation conditions, it should be emphasized that most buried pipes installed in the USA are not subject to prolonged UV or heat exposure. Therefore, these exposure elements do not play a major role in determining the longevity of PVC pipes. It is only when pipes are inadequately stored or have inadequate stabilizer added during processing that these degradation mechanisms can reduce the expected life of PVC pipes.

Current Performance

Responses from the performance survey indicate that PVC pipes are used to various extents within water utilities. Of those water utilities that supplied failure data, a range of annual failure rates (number/100miles(km)/year) were estimated. At the top of this range, South East Water (Australia) exhibited an estimated annual failure rate of 10.9 failures/100miles/year (6.8

failures/100km/year). At the bottom of the range, Epcor (Canada) exhibited an estimated annual failure rate of 0.3 failures/100miles/year (0.04 failures/100km/year). An examination of the UK nationally agreed database of water mains failures indicates an average failure rate of 11.7 failures/100miles/year (7.3 failures/100km/year) for UK PVC pipes. Because of variations in the methods of reporting data and the difficulties in cross-correlation, it is recommended that an effective method of recording and organizing pipe asset and failure data be used, to capture this variation. Until data can be represented in this manner, care must be taken in interpreting and comparing failure rates. It should also be noted that in most cases the number of recorded failures in PVC pipes was low compared to other materials. Consequently, expecting the breakdown in failure types as reported by this survey to be representative of PVC pipes as they age may not be realistic. However, it is still informative to compare current trends in failure types with other water utilities.

Examining the failures sustained in individual water utilities illustrated the dominant failure modes in each case. Since it is likely that the external factors governing pipe failure (i.e., ground movement) acted in different magnitudes for each water authority, different dominant failure modes were observed in each case. In some, joint leakage or fittings and especially tapping appeared to be a problem, While for others the fraction of longitudinal and circumferential fractures was relatively high.

A set of samples from fracture failures in Australian PVC pipes has been examined. In most cases, crack initiation occurred at inherent defects in the pipe wall. These may be attributed to cavities from unfilled air bubbles during the pipe manufacturing process, or foreign particles (dust or grit) that were included during pipe extrusion. Crack initiation occurred in all cases examined at, or near to, the inside surface of the pipe and then propagated through the wall thickness, finally penetrating the pipe outer surface. Longitudinal fractures which were relatively short in length were observed in most cases. This may indicate that crack growth in the cases examined, occurred under normal operating conditions, in the absence of extreme loading events such as pressure surges, which have been known to propagate cracks for longer distances. Scanning electron microscope examination of fracture surfaces also indicated that crack growth was usually accompanied by local ductility and the resulting fracture was not purely brittle.

The variability in failure modes from authority to authority indicates that installation practices may be a much larger factor in determining the types of failure modes, than pipe manufacture. Consequently, the collection of failure data and detailed analysis of the data using consistent formats could lead to early detection of factors affecting premature failure of PVC. However, an examination of failure surfaces from the limited number of collected pipes indicated that, of the failures examined, manufacturing defects played a role in premature failure.

Test Procedures Used for Predicting Lifetimes

The performance of PVC-U, PVC-M and PVC-O formulations sourced from the USA and Australia was assessed using a range of tests. These tests concentrated on assessing the performance of PVC for brittle and ductile failure. In the area of resistance to brittle failure, the C-ring test was used to determine the variation in time to failure with initial applied stress intensity factor (SIF). For PVC-U samples, tin-stabilized pipe manufactured in the USA exhibited the longest failure times for a given initial applied SIF, which was followed by Ca/Zn-stabilized pipe from Australia, lead-stabilized pipe from Australia, and finally tin-stabilized pipe from Australia. However, when the C-ring data was extrapolated back to estimate plane-strain fracture toughness values (at a failure time of 15 minutes), all formulations passed the performance requirement stipulated in the Australia/New Zealand Standard AS/NZS 1477 (Standards Australia/Standards New Zealand 1999) which stipulates minimum acceptable fracture toughness levels for PVC-U pipes. When PVC-M was assessed with this same test, examination of fracture surfaces of PVC-M C-ring samples revealed that failure occurred by ductile tearing rather than brittle fracture. Consequently, the use of an applied SIF to characterize failure for ductile pipes such as PVC-M is questionable.

The superior performance of PVC-U Tin (USA) pipe in the C-ring test was attributed to an increase in crack incubation times and a decrease in the slow crack growth rate. However, it should be noted that only one pipe manufacturer's material was assessed in this test, and the performance of this pipe may not be indicative of the performance of all manufacturers. Variations in performance are to be expected, thus highlighting the need for suitable performance requirements to be included in the manufacturing standard.

For more ductile pipes such as PVC-M and PVC-O, the C-ring test was not suitable and an essential work of fracture (EWF) test was used to assess performance; for comparison, this test was also performed on the PVC-U samples. In the EWF test the plane-stress fracture resistance (w_e) was used to characterize ductile fracture. As expected, the performance of the PVC-M pipes was significantly better than the PVC-U pipes, which reflects the influence of the ductility-promoting modifiers used in these materials. For PVC-M, the Ca/Zn-stabilized product appeared to be the more ductile. The EWF test shows promise for assessing the performance of ductile pipe materials such as PVC-M, however, for pipes such as PVC-U which fail in a brittle manner in the field, the C-ring test is considered more applicable. For PVC-O, EWF results indicate that the value of w_e perpendicular to the molecular alignment in these pipes is much greater than the w_e value parallel to alignment. This is explained because crack propagation (and the creation of new surface area) perpendicular to molecular alignment requires molecular chain scission to occur and, consequently, a greater amount of external work. In contrast, crack growth parallel to the direction of molecular alignment requires only the separation of secondary bonds between molecules. This difference in w_e is reflected in C-ring tests on PVC-O, where crack propagation occurs in the circumferential direction rather than in the radial direction.

Lifetime Prediction

A fracture mechanics failure model has been developed to predict failure times in PVC-U pipes under combined internal pressure and deflection loads. The method combines time-dependant fracture toughness, slow crack growth rate and individual applied SIFs for internal pressure, diametrical deflection and residual stress loading. The fracture mechanics failure model was compared against experimental simulation of in-service loading, in which an internally notched pipe was pressurized and subjected to parallel plate deflection loads. The fracture mechanics model was used to predict an incubation period and a period of slow crack growth to failure. Failure was predicted to occur at a critical crack length when the applied SIF exceeded the material fracture toughness. The model predictions agree well with recorded failure times under the range of experimentally applied SIFs. While this agreement is encouraging, the applicability of the incubation periods for pipes in service requires further validation as there is evidence that they may not exist where sharp notches exist in the pipe wall. Furthermore, although good agreement was obtained for PVC-U pipes, it should be noted that Linear Elastic

Fracture Mechanics theory cannot be used to predict failure in PVC-M and PVC-O, which do not exhibit brittle fracture failure modes.

Predicting Network Performance

To capture uncertainty in the pipe fracture process, a Monte Carlo simulation model has been developed to predict the performance of PVC pipelines. Based on the fracture mechanics failure model developed, the Monte Carlo simulation generates a hypothetical pipeline comprised of a series of independent pipe segments. The maximum inherent defect size in each pipe segment is treated as a stochastic variable, which (based on an examination of fracture surfaces from field failures) is represented by the Weibull probability distribution function.

For each segment in the hypothetical pipeline, the Monte Carlo simulation randomly assigns a maximum inherent defect size, based on the Weibull distribution. A time-marching loop is then performed in which incremental crack growth is simulated in each pipe segment. Failure is assumed to occur by brittle fracture when the applied SIF exceeds the material fracture toughness. An alternative criterion for ductile fracture is satisfied when the applied net section stress exceeds the material yield strength. Upon failure, a new pipe segment is installed and an inherent defect assigned. Output from the Monte Carlo simulation provides the number of new pipe segments installed for each year of the simulation, and the failure years for each pipe segment. Combining this data allows the average failure rate/pipe length/year) to be estimated.

To derive simpler expressions for average failure rate as a function of pipe age, lifetimes generated from the Monte Carlo simulation were represented using the Weibull probability distribution. Making the simplifying assumption that the number of failures along the pipeline followed a binomial process, allowed the expected number of failures to be described by the Weibull hazard (or failure rate) function. Finally, the Weibull hazard function for pipe lifetime was described in terms of operating pressure, for a prescribed pipe size and external loading conditions.

To account for the effect of pipe size on average failure rate, the Weibull distribution parameters for inherent defect size were scaled to the assessed pipe by using the ratio between the internal pipe surfaces of a DN100 pipe and the assessed pipe. After repeating Monte Carlo simulations with these new defect distribution parameters, a set of simple equations were proposed that capture the effect of pipe size and operating pressure on average failure rate over time. However, it should be noted that these equations should be restricted to the following pipe sizes:

- Class 150 psi PVC-U (Tin) (USA) pipes, DN 4 inch to DN 12 inch
- Class 1.2 MPa, Series 2 PVC-U (Ca/Zn) (AUS) pipes, DN 100 mm to DN 300 mm

A simple set of "look up" graphs were also compiled for these ranges and are given in Appendix E of this report. Outside of these ranges individual calculations need to be calculated for each pipe and a representative defect distribution calculated. From a comparison with the UK Water Industry Research (UKWIR) national database of water mains failures, reasonable agreement was obtained between actual and predicted failure rates for pipes of different ages.

Standardization

The current standards under which PVC pipes are manufactured and installed in North America were assessed in detail and a significant number of improvements are proposed. Specifically:

- The performance requirements on resin may be over prescriptive and lead to excessive test requirements, limit manufacturer innovation, and will not guarantee a high-performance PVC pipe. It is suggested that resin selection be left to the pipe manufacturer and that adequate performance requirements be included in the pipe manufacturing standard.
- The current method for calculating the hydrostatic design basis (HDB) for PVC pressure pipes does not take data scatter into full account. Consideration should be given to adopting the ISO 9080 practice for HDB calculation where the lower 97.5% confidence limit is used.
- Design stresses within AWWA and also with ASTM standards are not consistent and should be revised for consistency. The safety factors that define the final design stress should be based on a more scientific basis than those currently used. The current design stresses used within the United States for PVC-U and PVC-M are generally lower than those used elsewhere in the world, and require manufacturers to use more resin for similar performance.
- The practice of incorporating a surge allowance into the pressure class designation for PVC pipe is questionable and for most pipelines operating under static pressures this is not needed. For pipes operating under surge or fatigue conditions, the application of a surge allowance is overly simplistic. Such pipelines should be specifically designed for these conditions.
- Current short-term testing is prescriptive, does not confirm long-term pressure performance, and does not consider resin variation. It is recommended that short-term pressure testing to confirm the long-term HDB be adopted. This testing should allow for changes in resin selection and a suggested test procedure uses the 99.5% lower prediction limit value at one hour determined from the 50-year linear regression line for the appropriate resin, determined using ISO 9080.
- Field failures invariably occur via a slow crack growth mechanism, rather than the ductile mechanism currently predicted by the standards. Consideration should be given to include a fracture toughness performance requirement in North American PVC-U pipe standards, as well as a test to ensure adequate gelation for PVC-U pipes, and a ductility test for PVC-M pipes. *(Note: PVC-M pipes are not being manufactured or sold in the United States.)*. For PVC-U resins the current acetone test may be too benign and consequently may not ensure adequate gelation levels. If a solvent test is still considered the most suitable method for determining gelation, the solvent should be changed to dichloromethane. Where occupational health and safety is an issue, another test method such as differential scanning calorimetry should be considered, however no suitable test method has so far been developed, although an ISO study is in progress. The Australian industry is currently trialing an optical method that may be satisfactory for the United States industry.

- Current United States standards do not specify any performance requirements for either thermal or UV stabilization. These factors may not be an issue for many parts of the United States, however, for southern climates where heat and UV may affect the longer term performance, consideration should be given to these two factors. Whilst prescriptive factors such as defining a specific level of titanium dioxide (TiO_2) could be satisfactory, a better solution may be to define a minimum oxygen and dehydrochlorination induction time (OIT & DIT), This would ensure that adequate stabilizers had been incorporated into the PVC resin. An alternative would be to define minimum changes in pipe properties when the material has been exposed to a weathering regime.
- The impact requirements of USA standards are significantly higher than those required elsewhere in the world. These impact requirements may have beneficial effects in that they require the PVC manufacturing processes used in the United States to produce more ductile pipes. However, this needs more investigation. It is recommended that the USA industry undertake a benchmarking study to determine the reasons for the higher ductility of the Sn-stabilized pipe assessed in this study, and determine if it is representative of standard manufacturing practices.
- Both a high-temperature oven test and a reversion test should be considered for inclusion into USA standards to ensure a high quality of manufacture.
- Standards for ensuring the quality of materials for use in contact with drinking water are considered to be suitable to ensure that the pipe material does not contaminate potable water.
- Consistency does not exist between AWWA and ASTM standards in the United States. A case exists for the development of one consistent standard for each of PVC-U, PVC-M, and PVC-O.
- It is recommended that the industry in the United States develops national training courses for the jointing and installation of PVC pipes, and that attendance at these training courses become mandatory for any staff involved in the installation of PVC pipe by a water authority. Strategies for supervision of PVC pipe installation should also be developed by the industry, and these should be adopted as industry best practice.

FUTURE RESEARCH

In developing models to predict PVC pipe lifetimes, a significant amount of information is needed. In some cases, only a limited amount of information was available and realistic estimates of input parameters have been assumed. It is recommended that additional information be generated in the following areas to increase the accuracy of failure rate predictions from the models.

Inherent Defect Size

Only fracture failures that occurred in Australian PVC-U pipe were examined in this study. Consequently, the applicability of the Weibull distribution for inherent defect size to USA/Canadian PVC pipes is not yet validated. To this end, pipe samples from USA/Canadian field failures should be exhumed and the fracture surfaces examined. While an initial examination under optical microscope will reveal the origin of crack initiation, scanning electron microscopy should be used to quantify the inherent defect size.

It should also be noted that defect size data in this study were obtained from a range of pipe sizes. If these inherent defects are created during the pipe manufacturing process (cavities, for example), their size distribution may also depend on pipe diameter and wall thickness. To this end, a continued program of field failure examination is required to determine the influence of pipe size on inherent defect size.

Applied Stress Intensity Factors for Different Defect Shapes and Locations

The failure model developed in this study describes SIFs in terms of radial defect size only. To provide a "worst-case scenario", defects have also been assumed to be located at the pipe inner surface, where circumferential stresses from internal pressure, diametrical deflection, and residual stress are highest. In practice, a range of defect shapes and pipe wall locations will exist and SIF solutions are available (Rooke and Cartwright 1976). However, while these improvements to the physical model could be readily incorporated into the Monte Carlo simulation, there is currently insufficient data to describe probability distributions for defect size and shape. A continued program of field failure examination is required to solve this limitation.

Loading Conditions

The fracture mechanics failure model requires the applied circumferential stresses due to internal pressure, external loading, and residual stress as input. While internal pressure and residual stress may be well defined, external loading is idealized in the model as a soil weight and a surface load acting above a buried pipe. A more realistic model should account for external loads generated from frost (in cold climates) and expansive clays (due to moisture depletion).

Crack Incubation Times for Pipes in Service

For pipes in service, the initial applied SIFs are relatively low, which gives unrealistically high incubation periods when predicted from fracture mechanics theory. While predicted incubation times are reasonable for higher applied SIFs, an alternative approach is required to address pipes in service. Following previous work conducted at CSIRO, an experimental program to observe incubation periods for a range of defect types and applied SIFs should be conducted. This will validate the current evidence suggesting that incubation periods do not exist for pipes operating under low applied SIFs in service.

Different Failure Modes

Although reasonable agreement was possible between the Monte Carlo simulation model and recorded failure data, it should be noted that the simulation described in this paper only accounts for those failures that occur by longitudinal fracture. Consequently, these longitudinal fracture failures should be isolated from the total recorded failure rate. Since failure rates in USA/Canadian water utilities are low, this may only be practical for Australian utilities. Although initial comparisons with longitudinal fracture data from Yarra Valley Water (Australia) are encouraging, the applicability of external loads used in the Monte Carlo simulation should be validated for the United States. This would, however, require the USA industry to adopt standard failure reporting procedures. Additional studies need to be carried out to predict the other failures that are commonly reported in PVC pipe networks, i.e. fittings.

Generic Failure Rate versus Age Curves

The current study has demonstrated that the Weibull distribution function is well suited to describing the variation in lifetimes, generated from the Monte Carlo simulation. However, it is reiterated that in the current study, particular pipe sizes and external loading conditions were chosen. For the models to be widely applied to water distribution networks with confidence, a benchmarking study needs to be undertaken to look at the effects of variability in pipe size, internal pressure, and external loading conditions on predicted failure rates. This would need to be done in conjunction with water utilities that have a large amount of reliable pipe failure data.

CHAPTER 1
LITERATURE REVIEW AND REPORT – WATER QUALITY IN PVC PIPES

INTRODUCTION

The presence of PVC pipes in water distribution systems has increased since the 1970s. In 1999, plastic pipes comprised 69% of the overall buried pipes used in water distribution worldwide, and PVC represented 66% of all buried pipe used for the transport of potable water (Rahman 2002). In North America, the representation is even higher: 95% of all small diameter pipe and 60% of pipes of diameters greater than 6 inches are made of PVC (Rahman 2002).

Historically, PVC pipe manufactured outside the USA and Canada has been stabilized with lead (Pb)-based salts. However, as concerns emerged about the migration of lead into drinking water and particularly into the environment after disposal, pipe manufacturers have curtailed their use of Pb-based stabilizer systems. In the USA and Canada, tin (Sn)-based compounds are currently the main stabilizer system used for PVC pressure pipes and sewer, while in Europe and Australia calcium/zinc (Ca/Zn) stabilizers have been adopted, and recently metal-free organic-based stabilizers (OBSs) have been introduced.

Despite the introduction of the new stabilizer systems, some water utilities still have apprehensions about the risk of stabilizer migration into water and the potential health hazards. This review aims to investigate those concerns and compare the methodologies available worldwide to ensure that materials used for contact with potable water are safe.

The consumption of PVC worldwide in pipes and fittings in 1999 totaled 7069 million tons. Asia consumed 32.1%, Europe 25.4%, North America 33.2% and South America 9.1% of this amount (Hruza, Chemson Pacific, pers. comm. 2002). A survey of stabilizer usage in that same year revealed that the predominant stabilizers used in pipes and fittings were Pb-based stabilizers (in most countries except the USA, Japan, Australia, New Zealand and some western Europe countries) (Table 1.1). This includes both sewer and drinking water pipes. The predominance of Pb-stabilized systems is attributed to their lower cost compared to Ca/Zn (at least 2% cheaper) and organic stabilizers. In Europe, despite the economics, the acceptance of Ca/Zn stabilizer systems has increased due to legislative pressure, associated with improvements in Ca/Zn systems, and the increasing familiarity of compounders to their processing requirements (Defosse 2002). Meanwhile, in America, Sn stabilizers will continue to be the predominant stabilizer system in the market, due to its water quality maintenance track record, and because local conditions make it cheaper than the alternatives (Defosse 2002).

Table 1.1
Ban of lead stabilizers in pipes

Country	Year
Sweden	January 2002 (voluntary)
Japan	1993
UK	Under consideration
Netherlands	2003–2004 (voluntary)
Denmark	December 2003

OBSs, which were introduced in the late 1990s, are reported to offer processability and performance comparable to Pb-based stabilizers. Initial results suggest good performance, and their acceptance into the pipe market is growing (Pfaender et al. 1998). One of the advantages of OBSs is their compatibility with other stabilizer systems, e.g., Sn- and Ca/Zn-based stabilizers (Crompton Vinyl Additives 2001).

STABILIZER MIGRATION

In studies on the migration of stabilizer systems, the greatest emphasis has been placed on Pb-stabilized pipes because of environmental concerns (Al–Malack 2001, Burn and Schafer 1997). The migration of lead from PVC pipes has been reviewed extensively (Burn 1993, Burn and Schafer 1997, Al–Malack 2001, Packham 1971a), with the consensus being that the concentration of lead leached into water does not pose a health hazard to humans, provided proper flushing is conducted during the commissioning of new pipes.

The large majority of papers available on organotins are on a different family of organotins than the ones used in the manufacture of PVC pipe. These papers were dedicated to studying the effect of highly substituted organotin species on the environment, where they can have detrimental effects on aquatic and other marine life forms. Poisoning of humans has to date been recorded only amongst workers handling organotin-based pesticides and fungicides during manufacture, i.e., at significantly higher concentrations than those present in water. Contamination or accumulation of organotins in humans has been rarely reported, partly because its toxicity to humans has not been proven (WHO 1993, NHMRC 1996), and partly because of the low levels detected in humans (Cooke 2002, Whalen and Longanathan 2001, Sadiki et al. 1996), where it has been reported the main source of organotins is attributed to the consumption of contaminated seafood. Again, it should be emphasized that this was from organotins formulated to kill organisms: pesticides, fungicides, and marine anti-fouling agents – different types of organotins to those used in making PVC pipe.

Information on the migration of other metal-based stabilizers is limited, partly because of the fact that Ca, Zn and insoluble barium (Ba) are not considered harmful, and no deleterious effects have been verified due to their ingestion.

OBSs, which are metal free, are advocated as having even better immobilization in the PVC molecule compared to the other metal-based stabilizers and hence lower migration rates. The research conducted to date has not detected deleterious health effects (Pfaender et al. 1998; Crompton Vinyl Additives 2001; Hruza, Chemson Pacific, pers. comm. 2002).

A study conducted by Al–Malack (2001) into the migration of Pb, Sn, cadmium (Cd), Ca, and Ba stabilizers from water pipes exposed to different temperatures, water characteristics and UV revealed that:

- In all stabilizer systems, the heavy metal migrates into water.
- The migration rate order was Ca > Pb > Sn > Ba > Cd; however, due to toxicity, Pb is more hazardous and Cd has been banned from pipe formulations due to toxicity.

Effect of Solvent Cement on Stabilizer Migration

The mobility of stabilizers is influenced by the processing methodology and by the selection of the stabilizers used in the system.

A survey of 17 American pipe manufacturers using various tin stabilizers was conducted in 1982 and was followed by stabilizer migration studies (Boettner et al. 1982). It revealed that the migration rate from PVC pipes was dependant on the manufacturer. This can be explained by variations in formulation (concentration and type of stabilizer), and by the processing conditions, as migration is strongly influenced by excess stabilizer and by degraded stabilizer present in the pipe.

Moreover, studies of leaching rates of tin stabilizers conducted by Boettner et al. (1982) and Al–Malack (2001) demonstrated that different stabilizer combinations have different leaching rates. For example, Boettner et al. (1982) verified that for pipes containing a combination of tin stabilizers, the leachate distribution over time would result in a bimodal pattern where different stabilizer types migrate at different rates, extending the leaching period.

EFFECT OF WATER QUALITY ON STABILIZER MIGRATION

Effect of Water Temperature on Stabilizer Migration

The rate of stabilizer migration increases with temperature. Analysis of the migration of tin stabilizers from commercial chlorinated poly (vinyl) chloride (CPVC) pipes at 24°C (75F) and 65°C (149F) showed a marked increase in tin extraction at the higher temperature (Forsyth and Jay 1997). Similar observations had been made for tin stabilizers (Forsyth and Jay 1997) and for lead stabilizers in PVC pipe (Burn and Sullivan 1993, Burn and Schafer 1997). The susceptibility of a metal stabilizer to temperature is dependant on the type of stabilizer used. A study conducted by Al–Malack (2001) verified that an increase in water temperature from 35°C (95°F) to 45°C (113°F) caused an increase in the tin concentration in water of 42%, as well as Ca, Sn, and Ba, as shown in Figure 1.1, however the migration of Pb and Cd was unaffected in that temperature range.

For tin stabilizers, this increased mobility can be explained by the accelerated degradation of the organotin stabilizer into lower substituted species, as discussed by Forsyth and Jay (1997) who studied the effect of increasing the temperature from ambient to 65°C (149°F) in CPVC.

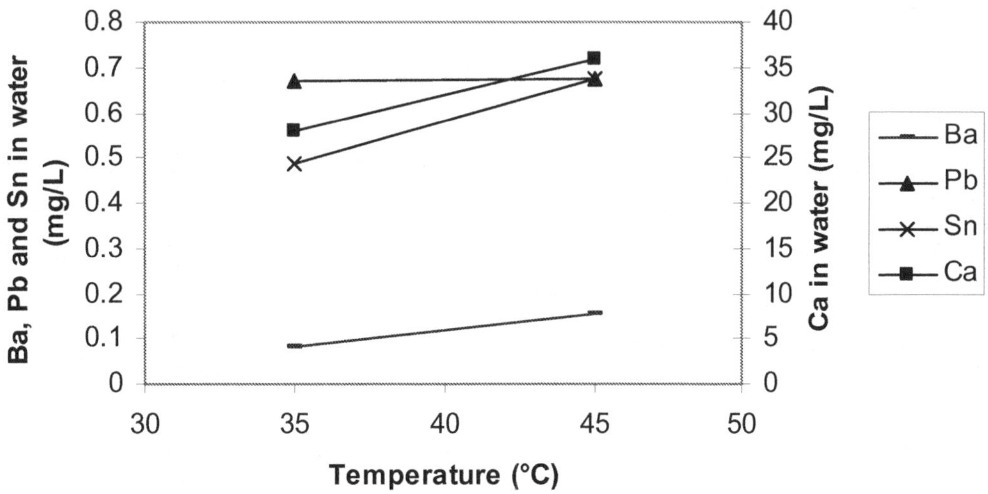

Figure 1.1 Example of extraction levels detected at 35°C and 45°C after a single static extraction over time (Data from Al–Malack 2001)

UV was also found to be an influencing factor in extraction of heavy metals; Al–Malack (2001) verified an increase in migration for pipes exposed to an artificial UV source. The increase in the lead and tin concentration was eight times more than the maximum contaminant level of lead in drinking water (0.115 mg/L after 12-hour exposure). Such an effect was most likely caused by the increase in pipe temperature upon exposure to the sun.

Effect of Water pH on Stabilizer Migration

The migration of stabilizers from PVC pipe is favored by acid pH, as demonstrated by Al–Malack (2001) (Figure 1.2) and Burn and Sullivan (1993) for Pb and for Sn, Ba, Ca, and Cd.

Effect of Total Dissolved Solids in Water on Stabilizer Migration

A higher concentration of total dissolved solids (TDS) increased the migration of Pb, Sn, Ba, Ca, and Cd stabilizers from the pipe (Al–Malack 2001). However, the rate of migration differs for each stabilizer. Tin in particular was observed to migrate at a more accelerated rate in the presence of a higher TDS concentration (Al–Malack 2001) (Figure 1.3). In other words, the presence of anions would appear to favor the mobility of the metal stabilizer.

EFFECT OF RETENTION TIME ON STABILIZER MIGRATION

Long retention times can be detrimental to the quality of water as it allows the amount of stabilizer migrating from the pipe to accumulate in the water (Figure 1.4). Long retention times also cause a number of other water quality concerns, one of which is the loss of the residual disinfectant used to maintain water quality.

Al-Malack (2001) showed that the accumulation of lead in water left stagnant for 10 hours resulted in levels in excess of the allowable USA Environmental Protection Agency (USEPA) concentration (0.01 mg/L). The lead level detected in that experiment was 1 mg/L at 2600 mg/L TDS. However, it also appears that the pipes used in that study had not been flushed before the experiment was initiated.

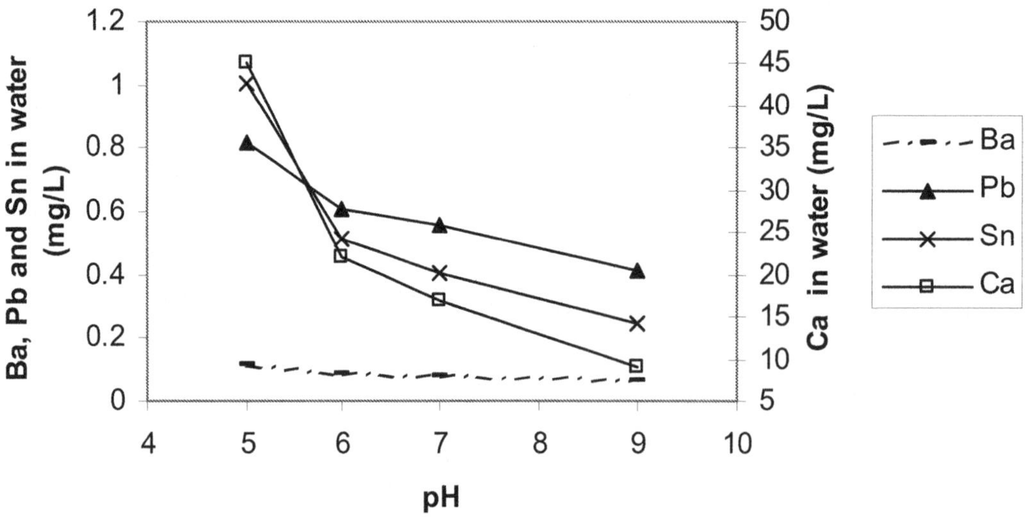

Figure 1.2 Effect of pH on the migration of metal stabilizers (Data from Al–Malack 2001)

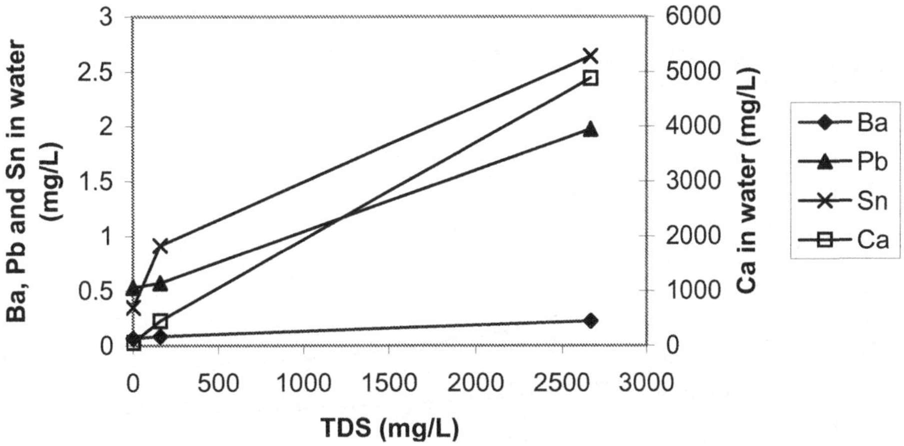

Figure 1.3 Migration rates for metal stabilizers in water (Data from Al–Malack 2001)

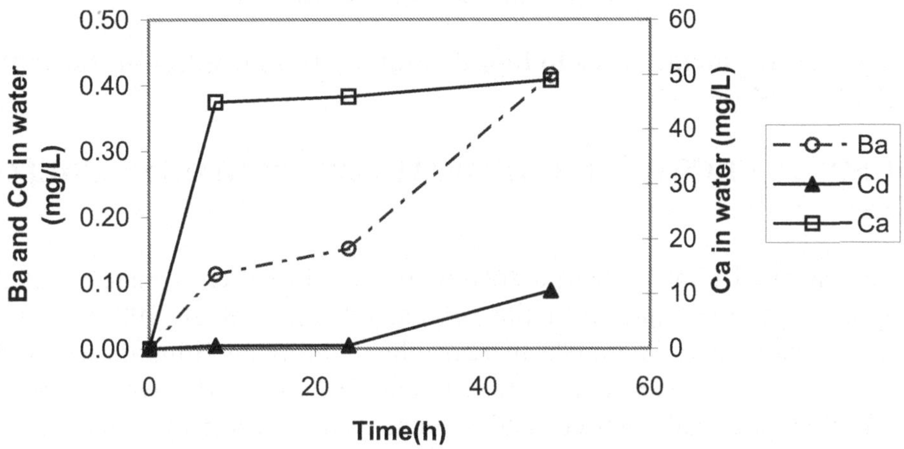

Figure 1.4 Stabilizer accumulation in stagnant water (Data from Al–Malack 2001)

Conversely, Forsyth and Jay (1997) demonstrated in experiments that the total amount of stabilizer that could be leached was higher using moving water extraction compared to stagnant water.

The pattern of leaching and the rate of migration decrease with operation time has been highlighted by Burn and Schaffer (1997), Sadiki and Williams (1999), and Boettner et al. (1982). In most cases, the reduction was drastic after the initial 24 hours of operation, although some studies detected migration up to 5 and 21 days, but at lower rates (Forsyth and Jay 1997).

EFFECT OF SURFACE TO VOLUME RATIO ON STABILIZER MIGRATION

Stabilizer migration is more noticeable in small diameter pipe, as the surface to volume ratio (S:V) is higher, and partly because of longer residence times of water in those pipes (Burn and Schafer 1997) (Figure 1.5). However, simply using larger diameter pipes in a distribution system is not necessarily the answer to minimizing the effects of stabilizer migration because of the increased water retention times and the corresponding decrease in water quality due to these longer retention times.

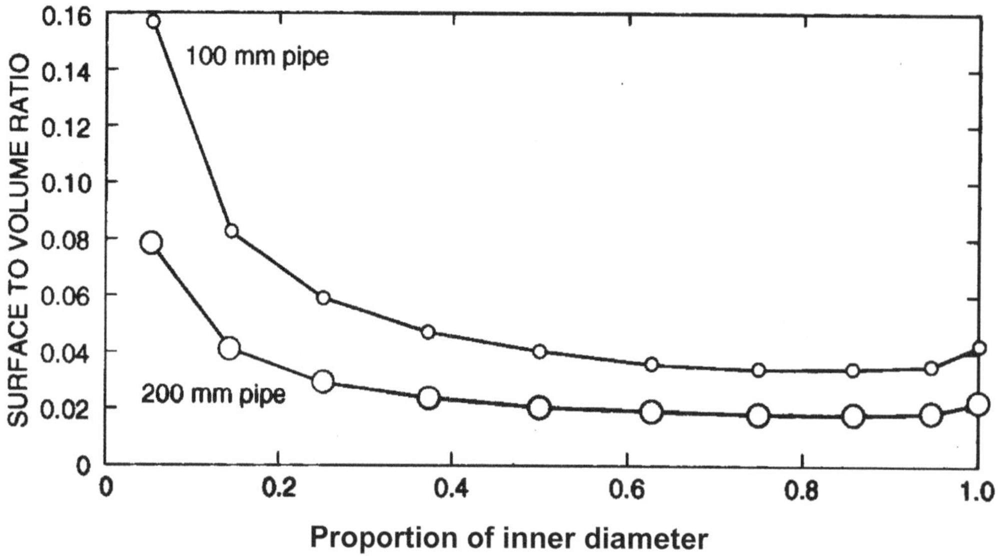

Figure 1.5 Effect of wet surface area to liquid volume ratio for different flows (Data from Burn and Schafer 1997)

EFFECT OF SOLVENT CEMENT (CHEMICAL JOINTING) ON STABILIZER MIGRATION

Solvent-cemented joints, where a solvent is used to soften a material at the joint, can cause contamination of the water with the solvent (Boettner et al. 1982), particularly when excess solvent cement is used; and could increase the mobility and migration of stabilizers from the joint, as suggested by Jones–Lepp, Varner, and Hilton (2001), who compared stabilizer migration from pipes that had been cut and solvent cemented together, and pipes that had no joints.

The increase in mobility of the stabilizer could have resulted from diffusion of the solvent to the surface during curing. Additionally, stabilizer contained in the solvent cement could also have migrated into the water, contributing to the increase in concentration (Jones-Lepp, Varner, and Hilton 2001).

TOXICITY OF STABILIZER SYSTEMS IN PVC PIPES

The toxic effects of lead are known and well documented (WHO 1996). The maximum concentration allowed is 10 µg/L according to the World Health Organisation (WHO), the European Committee for Standardization, and Standards Australia. Information on other stabilizer systems is more limited. Calcium, zinc, and tin in their inorganic forms are not considered hazardous. While some forms of organic tins can be toxic, the tin stabilizers used in PVC pipes exhibit low toxicity and are not considered to be an issue in relation to their migration into water.

Organotin Stabilizers

In PVC formulations, mainly tin mercaptides are used for heat stabilizers and tin maleates are used for UV stabilizers. For heat stabilizers, typically monomethyl tin, monobutyl tin, dimethyl tin, and octyltins are used. However, for PVC pipe stabilizers mono- and di-substituted tins tend to be the most common.

The migration of organotin stabilizers from PVC and CPVC pipes into water has been observed in laboratory experiments and in water distribution systems (WHO 1996, Sadiki and Williams 1999, Sadiki et al. 1996, Forsyth and Jay 1997, Boettner et al. 1982, Hoch 2001). Table 1.2 provides an indication of the range of concentrations of organotins detected in potable water and wastewater. Note that wastewater generally includes organotins from sources other than pipes.

Toxicity of Organotins

Most information available on organotins pertains to the more highly substituted organotin systems, e.g., tributyl tin which is an endocrine disruptor and is used in anti-fouling agents, pesticides and fungicides (Hoch 2001). These are not relevant to the manufacture of PVC pipe and information on the toxicity of non-pesticide organotin compounds used in PVC pipe is more limited (Sadiki et al. 1996, Boraiko et al. 2000).

Dialkyl tins have deleterious effects at high dosages, for example, dosages of 6 mg/kg body weight resulted in an increase in thymic and malignant tumours in rats, and dibutyl tin chloride appears to decrease food intake, reduce growth, induce anaemia, and immunodeficiency on rats and mice (WHO 1996).

However, the limited evidence available to date cannot confirm that di- and monobutyl tins are toxic to human. In-vitro studies verified that dibutyl tin at 74 □Mole can decrease the activity of the human enzyme aromatase, an enzyme involved in reproduction (Whalen and Longanathan 2001, Cooke 2002), and that it acts as a neurotoxicant at concentrations of 0.1 □Mole/L (Eskes et al. 1999). However, the in-vivo toxicity would also depend on the ease of transfer of the toxin from the blood to other organs and cells, and on metabolic processes, a factor that requires further investigation (Eskes et al. 1999, ORTEPA 1997). Other reviews conducted in Germany and in the USA suggest that organotin stabilizers used in PVC applications are safe (Summer et al. 1996 in ORTEPA 1997, McLellan and Bestervelt 1999 in Batt 2002).

Table 1.2
Organotin concentration in distributed water and wastewater

Type	Location	Range (ng/L)	Organotin	Reference
Drinking water	Canada	0.5–257 0.5–6.5	Methyl tin Dimethyl tin	Sadiki et al. (1996)
Drinking water	Florida, USA	0.49–8.1 0.4–2.2	Methyl tin Dimethyl tin	Braman and Tomkins in Sadiki et al. (1996)
Wastewater	Zurich, Switzerland	140–560 20–70 60–220	Monobutyl tin Dibutyl tin Tributyl tin	Fent (1996)
Wastewater	Bordeaux, France	20–70	Methyl, dibutyl and tributyl tin	Fent (1996)

A limited number of studies are available on the level of organotins encountered in humans (Table 1.3). The major source of organotins found in humans is likely to be from the consumption of seafood, with the contribution from PVC water distribution pipes deemed minor (Ebdon, Hill, and Rivas 1998). Higher organotins, such as tributyl tin which is used in marine anti-fouling products, degrade into mono- and dibutyl tin, which explains their presence in seafood, e.g., bivalves and crustaceans, and further up in the food chain. This explains the higher organotin concentration verified in Japan, where seafood constitutes a large part of the population's diet.

Pennikins (in Cooke 2002) proposed a tolerable daily intake of organotins of approximately 250 ng of tributyl tin/kg of body weight, based on toxicological studies. While the Scientific Committee for Food from the European Union has defined its own migration limits based on the tolerable daily intake levels for specific organotin species (Table 1.4) (ORTEPA 1997).

Table 1.3
Organotin concentrations detected in humans

Country	Concentration of organotin	Type of organotin	Location	Estimated organotin amount (mg/person)	Reference
Japan	14–22 ng/g 45–78 ng/g 2 ng/g	Methyl butyl tin Dibutyl tin Tributyl tin	Liver	153*	Takahashi et al. (1999) in Nielsen and Strand (2002)
Poland	2.4–11 ng/g	Butyl tins	Liver	16.5*	Kannan and Falandysz (1997) in Nielsen and Strand (2002)
USA	8.7 ng/mL 4.3 ng/mL 4.6 ng/mL	Methyl butyl tin Dibutyl tin Tributyl tin	Blood	0.105†	Kannan (1999) in Hoch (2001)

* Estimate based on the average adult human liver mass of 1.5 kg.
† Estimate based on an average adult blood volume of 6 L.

Table 1.4
European tolerable daily intake values for some common organotins

Organotin	Tolerable daily intake (μg/kg body weight/day)	Source
Dibutyl tin	5	Summer et al. (1996) in (Batt 2003)
Di-n-octyl tin	0.6	Commission of the European Communities – Scientific Committee for Food (1994) in Batt (2003)
Dimethyltin S"-bis	3	Commission of the European Communities – Synoptic Document No. 7, *Draft of Provisional List of Monomers and Additives Used in the Manufacture of Plastics and Coatings Intended to Come Into Contact With Foodstuffs*
Monomethyl tin tris	3	

ORGANOTIN MIGRATION

Mono- and dialkyl tins are added to unplasticised PVC (PVC-U) pipe formulations at levels between 10 and 20 g/kg of PVC (Hoch 2001). According to a survey by Mesch and Kugele (1992), the concentration that is leached from the distribution system is usually so low that it is not detectable, although the detection of tin stabilizer compounds in water was hampered for many years by the difficulty in speciation and separation of the different tin compounds due to limitations in the sensitivity of analytical techniques at the time.

In the mid-1980s, NSF International conducted stabilizer leaching studies on potable water pipes according to ANSI/NSF 14 (ANSI/NSF 2001) protocol. The study, conducted on 85 pipes containing twice the maximum allowed organotin levels, showed that 96% of the leachate samples had tin levels below the detectable limit, and the remaining 4% yielded a maximum of 25 μg/L of tin, which was less than half the NSF level of 50 μg/L allowed for tin at the time. Under ANSI/NSF 61 (ANSI/NSF 2002), tin is extracted at level I, meaning that it is found at concentrations of less than 10 μg/L (Mesch and Kugele 1992).

Similarly as with lead stabilizers, the concentration of organotin that leaches into water decreases with time (Forsyth and Jay 1997, Boettner et al. 1982), because the migration of tin into water is attributed to excess and degraded stabilizer located close to the surface of the polymer. Boettner (Boettner et al. 1982) showed that after installation, organotin levels decreased with the time of pipeline operation, even though organotin leaching could still be detected after 21 days under normal operating conditions. The long periods of leaching might be caused by differences in migration rates of the different organotin species contained in the pipe formulation (Boettner et al. 1982). Because of this, organotin levels can be significant in pipe sections where water remains stagnant for long periods of time.

Other researchers (Jones–Lepp, Varner and Hilton 2001) verified that stabilizer migration is faster in continuously flowing water compared to stagnant water. In flowing water Jones–Lepp, Varner and Hilton (2001) recorded concentrations of 1–5 μg/L, while in stagnant water the concentration was a maximum of 0.8 μg/L after four days. Quevauviller (1991); using PVC chips, verified a maximum of 0.2 μg/L of total tin extracted after 30 days. These facts highlight the importance of S:V for migration processes, and the effect that moving water can have on the diffusion of the stabilizer into water.

Effect of the Environment on Organotin Stabilizer Migration

Organotins degrade into lower substituted species under sunlight (Navio et al. 1993) and are digested by bacteria in aquatic environments through methylation and desalkylation, e.g., tributyl tin degrades into dibutyl tin and monomethyl tin (Forsyth and Jay 1997).

Effect of Solvent Cement on Organotin Stabilizer Migration

The use of solvent cements facilitates the leaching of organotin stabilizers, as shown by evidence from Jones–Lepp, Varner and Hilton (2001). Furthermore, Boettner et al. (1982) verified that certain volatile organic solvents (methyl ethyl ketone, tetrahydrofuran) from solvent cement used in PVC and CPVC pipe could leach into water for more than 14 days from installation in a laboratory pipe system, while cyclohexanone was only detected up to 1 day after installation (Boettner et al. 1982), thus the use of solvent cements with long cure times would facilitate leaching of organotins into water. Stabilizer concentrations detected ranged from 10 mg/L to 10 μg/L during the first 15 days of sampling. This would suggest that higher stabilizer migration rates would be expected from solvent-cement joints.

Speciation studies indicate that the low level of tin detected in water may be partly caused by adsorption of the organotin onto PVC, which also brings into question whether suspended matter in the water or biofilm growth would be able to adsorb tin stabilizer and reduce the amount of organotin in solution (Forsyth and Jay 1997).

For tributyl tin, pH influenced the solubility of the compound by promoting different ionization states, but it is not known if mono- and dibutyl tins can be affected in the same way. The pH was also verified to influence the speciation of inorganic tins as it facilitates their diffusion through the molecular walls of cyanobacteria. Most researchers agree that organotins can be extracted from water and that there is evidence that higher temperatures favor the migration process. However, there is still the need to determine the exact effect that pH, ionic water content, residence time, etc., have on the concentration of organotins and, most importantly, what effects the concentration of organotins has on the environment and human health (if any).

SAFETY ISSUES

Maximum Allowable Stabilizer Concentration

The broadest water regulatory body for water quality worldwide is the WHO. The WHO publishes guidelines for the maximum allowed concentrations for heavy metals and other hazardous substances that might find their way into drinking water (WHO 1996). These guidelines are accepted as the minimum standard, and are used by the majority of countries as the basis for their water quality guidelines. However, each country has the freedom to set the maximum concentration of each contaminant independently, according to directions from its national regulator. In the USA, the EPA is the regulatory body (USEPA 2002); in Europe the CEN (European Commission) sets the level, according to the European Drinking Water Directive 98/83/EC on the Quality of Water Intended for Human Consumption (Europa 1998), but individual member countries can choose to use more stringent limits; in Australia the levels are determined by the National Health and Medical Research Council (NHMRC) and the Australian Water Resources Council (NHMRC 1996). Comparatively, of all standards worldwide, the European contaminant levels tend to be the most stringent.

With regard to stabilizers, water quality standards, including the WHO guidelines, do not regulate the concentration of the stabilizer in water, but instead they regulate the concentration of heavy metals, such as lead, and other substances in water which can pose health risks. Tin stabilizers and OBSs are not regulated compounds and, in the case of tin, dialkyl tins are mentioned, but insufficient data is available to allow a maximum concentration to be established (Table 1.5). OBSs have only been used recently, but the limited data published to date do not indicate any health issues with their use (Hruza, Chemson Pacific, pers. comm. 2002).

Table 1.5
Examples of concentrations specified in the WHO guidelines that could be affected by stabilizer migration (WHO 1996)

Metal/substance	Maximum concentration (mg/L)
Cd	≤0.003
Pb	≤0.010*
Zn	Not a health hazard
Sn	Not a health hazard
Ca	Not a health hazard
Dialkyl tin	No limit established due to inadequate data
Tributyl tin oxide	0.002

* In the USA, the action level (maximum contaminant level) for lead is 0.015 mg/L, and the maximum contaminant level guideline is zero.

Test Methodology

Concern over potential health risks associated with the migration of substances from PVC pipes and other materials used in the water distribution system into water, has been reflected in standards for pipes used for transport of potable water (see Table 1.6). In the UK, pipe standards contain clauses that specify that the product should not be "non-toxic", it should not be organoleptic, and it should not support microbial growth.

Table 1.6
Main standards for analysis of substance migration into water

Standard	Title	Origin	Country of use
ANSI/NSF 61	Drinking water systems components – health effects	ANSI/NSF International	USA
AS/NZS 4020	Products for use in contact with drinking water	Standards Australia and Standards New Zealand	Australia and New Zealand
BS 6920	Suitability of non-metallic products for use in contact with water intended for human consumption with regard to their effect on the quality of water	British Standards Institute (BSI)	United Kingdom
ISO 3114	UPVC pipes for potable water supply: extractability of lead and tin – test method	International Organization for Standardization (ISO)	Worldwide
ISO 8795	Plastics piping systems for the transport of water intended for human consumption: migration assessment: determination of migration values for plastic pipes and fittings and their joints	ISO	Worldwide

Safeguards for water quality include, among other quality requirements, the assessment of migration from products used in contact with potable water, and the regulation of the concentration of adverse substances. For example, in the USA, potable water pipe standards require that the pipe or fitting is tested and certified according to ANSI/NSF 61, or as required by the health effect requirements in ANSI/NSF 14 for regulated substances.

Similarly, in Australia, compliance with AS/NZS 4020 (Standards Australia/Standards New Zealand 2000a) is required. European standards (EN) for pipes provide clauses alerting that the "material shall have no adverse effects on the quality of drinking water," and it refers to the need for approval and regulations that are not covered in the specific pipe standard before the material can be introduced into the market.

Evaluation of the effect of materials on water quality uses standards such as BS 3505 (BSI 1986) and AS 4020 (Standards Australia/Standards New Zealand 2000a), that require testing according to methods specified by the regulatory body, and to levels of contaminants specified by the regulator or by the WHO guidelines, whichever is most rigorous. This generally means compliance with BS 6920-1 (BSI 2000). However, any new products are required to be tested for specific chemicals, chlorine demand, organoleptic, and growth of micro-organisms (Franklin 2002), as outlined in ISO 8795 (ISO 2001), which is also a CEN standard.

Standards covering methods to determine the migration of specific substances from pipe are few. For example, ISO 3114 (ISO 1977) is tailored to the migration and the detection of lead salts and organic derivations of tin. BS 6920-1 (BSI 2000) is for migration from non-metallic pipes, While the two ANSI/NSF (2001, 2002) tests are aimed at the migration of any substance that might be of concern.

METHODOLOGIES USED FOR MIGRATION EVALUATION

BS 6920-1

This standard determines the suitability of non-metallic products for use in contact with water intended for human consumption, with regard to their effect on the quality of water. It is comprised of four parts: specification, methods of test, high temperature, and gas chromatography/mass spectroscopy identification of water leachate organic substances. This standard assesses a product for its effect on odor, flavor, appearance, micro-organism growth, cytotoxicity, and metal extraction.

ANSI/NSF 14 and ANSI/NSF 61

These standards are issued by the American National Standards Institute and NSF International. ANSI/NSF 14 covers performance requirements for plastic piping system components. It requires that plastic piping systems and components for use with potable water comply with ANSI/NSF 61. ANSI/NSF 61 covers procedures and the migration of indirect additives, products, and materials into water, and requires that PVC and CPVC pipes are tested for regulated metals, tin, antimony, phenolics, VOCs, residual vinyl chloride monomer, $CaCO_3$, calcium stearate, and other formulation-dependant species. It does not require taste and odor determination, nor microbiological testing. The pH of the test water is selected according to the type of substance that needs to be extracted; in the case of metals, the recommended pH is 5 or 10. The test procedure has variations for hot and cold applications, i.e., 82°C (180°F) and 23°C

(73°F) respectively; and uses water with 2 mg/L active chlorine and 100 mg/L hardness. The test consists of preconditioning at the test temperature with stagnant water for 14 days (cold applications), and for 1 hour for hot applications, sample extraction is performed using stagnant water for allocated time periods.

The contaminant is classified into one of four different groups according to the concentration of the leachate in the water – the higher the concentration, the more complex the toxicity tests are required.

AS/NZS 4020

This standard was based on BS 6920, but was modified for specific Australian requirements. It includes taste, appearance, micro-organism growth, cytotoxicity, mutagenicity, metal extraction, and high temperature tests. Among the improvements are the requirement for area to volume ratio of the test piece corresponding at least to the actual application, modifications to the leaching of toxic substances, and the inclusion of testing at high temperature. Extraction is conducted with distilled water adjusted to 50 mg/L hardness and 1 mg/L Cl_2, if chlorination is required. Extraction is carried out at 20°C (68°F) or at the maximum temperature for which the product is intended. The procedure consists of pre-rinsing the product to remove debris and extraction using stagnant water, which is renewed after 24-hour periods. Samples are collected for analysis after the first and the seventh extraction. This allows comparison between the initial and the "in-service" migration rates for the pipe.

This standard is very broad and is aimed at any product or material that could be used in contact with drinking water, including metal, plastics, and concrete.

ISO 3114

This test method is specifically tailored to two types of stabilizers: lead salts, and organic derivations of tin (mainly dialkyl tin, butyl tin, and higher homologues). The migration is evaluated with stagnant water at pH 4.5 + 0.1 at 20°C (68F) for periods of 48 hours after the pipe has been prewashed with water of pH 7–8.

ISO 8795

This standard is intended for the evaluation of migration for plastic pipes and joints used for transport of water intended for human consumption. It includes test water of pH 7, 9 and 11 (pH 9 is representative of European water quality); water quality was specified at 1 mg/L Cl_2, 10 mS/m maximum conductivity and a total organic carbon (TOC) $\leq$ 0.2 mg/L. The test temperature considered a range depending on the application of the product: 23°C (73F), 27°C (81F) (for tropical climates), and 70°C (158F) and 90°C (194F) (for warm and hot water systems). The extraction procedure consisted of presoaking the sample for 24 hours at ambient temperature, and for an additional period at the test temperature for hot water systems, followed by rinsing in running water before sample collection and extraction after 72 hours in cold water or 24 hours in warm water (ISO 2001). The water from the initial presoak period is discarded, which allows the majority of the excess stabilizer to be removed before extraction measurements are determined. The standard also provides diagrams of the experimental set-up for large diameter pipes.

Comparison of Standards for Stabilizer Migration

The standard tests are based on migration into stagnant water for periods of 24 hours and, if the concentration is higher than the Drinking Water Guidelines, migration is evaluated for subsequent periods of 24 hours (e.g., totaling 10 days for AS/NZS 4020). Variations in the methodology employed include: use of full sized products or pieces of material – consideration of S:V; type of extractant used – chlorinated water, distilled water, pH; method of evaluation (retention interval); flushing procedure (number of flushes before and between sample collection for analysis); and the testing temperatures.

The Australian Standard is broad and encompasses all aspects of water quality, while the ANSI/NSF 61 is specific to the determination of the concentration of the extractant; the final concentration then determines the types of tests required (teratology, genetic toxicity, etc.). Such measures streamline the analytical process and eliminate unnecessary testing, as many substances are not hazardous below a certain concentration.

The quality of the water used for extraction is important (see Table 1.7) as it can affect the migration rate. A useful improvement of ANSI/NSF 61 over other standards is that the pH of the water used in the extraction is determined by the type of substance being analyzed. This is important as the solubility and diffusion of some substances is dependant on pH. In the case of metal-based heat stabilizers, acid pH increases the rate of stabilizer migration into water (Al–Malack 2001). It is also important to consider the pH of the water where the product will be used, as considered in ISO 8795. For example, water of pH 9 better reflects European conditions than neutral water.

The option of using chlorinated water in the extraction procedure is provided for in AS/NZS 4020, ISO 8795 and ANSI/NSF 61. The chlorine level specified in the latter is higher than that used in the European and Australian standards. The water hardness value in ANSI/NSF 61 is twice the value specified in AS/NZS 4020, which, depending on the type of stabilizer used, could result in an increase in the migration rate.

The temperature of the water in the procedure is particularly important as higher temperatures increase the migration rate, therefore it has to reflect the conditions that the product will endure during use. ISO 8795 uses a higher test temperature (27°C/ 81°F) for pipes destined for use in tropical climates, which could be recommended.

Table 1.7
Characteristics of extractant water used in some common standards

Standard	Temperature* (°C)	Hardness (mg/L)	pH	Residual chlorine (mg/L)	Conductivity (mS/m at 25°C)	TOC† (mg/L)
AS/NZS 4020 Method J	C 20 H 40–80	50	—	1 + 0.1	—	—
ANS/NSI I61	C 23 H 82	100	5, 8, 10	2	—	—
ISO 8795	C 23 or 27 W 70 H 90	—	4.5, 7, 9, 11	1	Max. 10	Max. 0.2

* C = cold water application; W = warm water application; H = hot water application.
† TOC = total organic carbon.

Conditioning the sample before extraction, as required in ANSI/NSF 61 and ISO 8795, aims to remove excess stabilizer from the surface of the pipe so that migration levels can reach the "typical" rates observed in service. In contrast, AS/NZS 4020 instead determines the initial concentration, and is consequently more stringent since the concentration of leachate must be regulated from the start of the test. The ANSI/NSF and ISO standards mirror pipe conditions that would occur if good installation practices have been followed, where the line is first flushed, then disinfected, and then placed in service, whereas the Australian standard simulates a worst-case scenario.

In conclusion, the practice outlined in AS/NZS 4020 is considered the most conservative, but it is advisable that the water quality of the area where the pipe will be used be considered in the experiment, so that the extractant water can be adjusted accordingly (e.g., pH, and hardness), and that a clause be set for verification of migration of solvent-cemented joints. For fittings and other connections which are joined with solvent cement, it is also advisable to verify the rate of migration of substances from the joint, as solvent cement has been reported to increase the rate of stabilizer migration (Jones–Lepp, Varner and Hilton 2001).

Alternatively, flushing with water of a lower pH during commissioning could help remove excess stabilizer from the pipe, however, the feasibility of such a process would have to be considered.

PROCEDURES FOR MITIGATION OF STABILIZER RESIDUES IN WATER

Commissioning of new PVC pipes for service includes procedures for determining leaks, cleaning, and disinfection, but not to specifically remove excess stabilizer. These procedures are generally set by each water authority and aim to verify the integrity of the joints, remove installation debris, and prevent bacterial contamination. Upon operation, the levels of stabilizer migrating from PVC-U pipes tend to decrease to trace levels (Packham 1971b, Sadiki and Williams 1999). However, the period of greatest stabilizer migration from the pipe into water occurs during the initial operation after installation, when excess and degraded stabilizer accumulated on the pipe surface due to processing is removed, and can result in metal concentrations in excess of the recommended guideline values. Such temporarily high concentrations can be minimized by preflushing the pipes during commissioning.

Burn and Sullivan (1993) compared a range of commissioning routines used by Australian water utilities on new PVC pipelines, including stagnant water, water replacement and continuous flow, and determined their effects on the migration of lead in the laboratory and in the field.

Laboratory trials confirmed that lead migration was higher under:

- Acid pH (Burn and Sullivan 1993, Packham 1971a).
- Flowing water; lead removal increased with flush rate, but reached a plateau between velocities of 4 and 600 m/h.
- When using batches of stagnant water, the rate of lead extraction was higher by conducting water changes at 1-hour intervals compared to shorter and more frequent 15-minute intervals.

The evaluation of field procedures compared pipe lengths of 55, 85, and 220 m on dead ends subjected to the conditions specified in Table 1.8. The lead extraction after each of the treatments is compared in Figure 1.6. Use of the pipe "as received" (without any pretreatment) resulted in the highest migration rate; in comparison, the replacement treatment, which incorporated swabbing and disinfection, resulted in a 50% reduction in extracted lead. The continuous flow treatment resulted in the highest extraction rates in the first couple of days, followed by a subsequent decline in extracted lead. However, the authors remarked that the reduction at seven days did not follow trends observed in the laboratory trials, and could have been affected by an extraneous factor (Burn and Sullivan 1993).

Table 1.8
Commissioning procedures used in field trials by Burn and Sullivan (1993)

Procedure	Stagnant flow	Continuous flow	Water replacement
Pipe length (m)	55	85	220
Pretreatment	None	Pipe filled 8-hour flush at 1.5 m/min	Pipe filled 15-min flush at 120–150 m/min 15-min stagnant water 3 foam swabs 1-hour stagnant disinfection with chlorinated water 1 foam swab 1-hour stagnant water
Stabilizer migration	Pipe filled with water	Pipe filled with water	Pipe filled with water
Sampling frequency	Daily over 14-day period	Daily over 14-day period	Daily over 14-day period

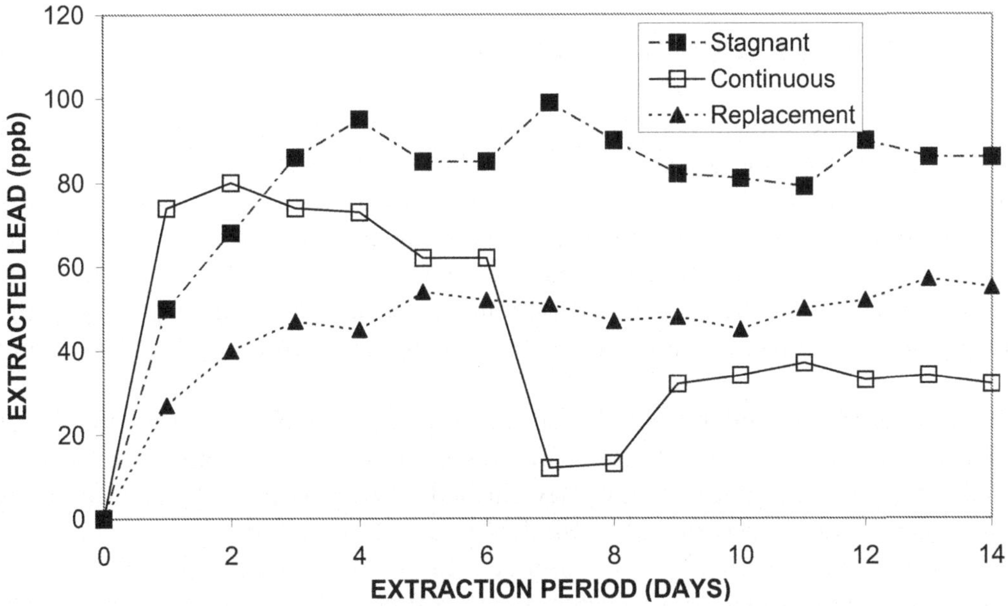

Figure 1.6 Lead extraction after commissioning procedures (Data from Burn 1993)

Their recommendation was that pipes could be subjected to an acidic prewash (with a weak acid) in the factory manufacture to remove most residual stabilizer; or to adopt either of two procedures used by the Rural Water Commission of Victoria. If a slow continuous flush procedure was required:

a. Main flushed and maintained at minimum flow conditions.
b. Day 1–3: 2 water changes per day.
c. Day 4–10: 1 water change per day.
d. Day 11–17: 0.5 water changes per day.
e. Day 18–24: 0.25 water changes per day.
f. Tappings allowed on day 1.

Alternatively:

a. Main flushed.
b. Main flushed 1 day after filling.
c. Main flushed again 1 week after filling.
d. Main flushed again 2 weeks after filling.
e. Main flushed again 3 weeks after filling.
f. Main flushed again 4 weeks after filling.
g. No tappings allowed before day 22.

Both procedures are, however, conservative. Alternatively, a water replacement procedure (i.e., a series of three water changes at 150-minute intervals, with turbulent water replacement between each water change, using a combination of air scouring, or swabbing and disinfection, followed by continuous low-rate flushing) is likely to reduce water usage and installation time. However, they did qualify this, saying the extent of its effectiveness would have to be confirmed.

CONCLUSIONS AND RECOMMENDATIONS

Evidence to date indicates that metal stabilizers migrate into water from manufactured PVC pipes. Research on the toxicity of the metals used in heat stabilizers has resulted in the setting of a maximum contaminant level of 0.010 mg/L for lead in Australia and Europe, and a target level of 0 mg/L in the USA (0.015 mg/L action level). Levels for organotin have not been established by the WHO and national regulatory bodies, due to insufficient evidence to confirm toxicity. As a precaution, the European Community adopts TLV values for some of the most common organotin stabilizers. Ca/Zn stabilizer systems are considered non-toxic and safe for use.

The literature into migration of lead and organotin stabilizers used in PVC pipes has shown that the concentration levels found in water due to PVC leaching are below the levels considered safe, provided pipes are properly flushed when first put into service, as the highest rate of stabilizer migration occurs in the initial 24 hours of operation.

The contribution of metal from stabilizers to environmental pollution is not considered significant compared to other sources such as air pollution. For lead stabilizers, mobility only becomes an issue during incineration of pipes and, in the case of organotins, bacteria in aquatic environments can break down organotin in solution. However, the rate of migration of stabilizers will depend on the manufacturing process and on the type of stabilizers used in the pipe. Therefore, it is necessary that new products be tested for stabilizer migration levels.

A number of environmental factors will affect the rate of stabilizer migration, e.g.:

- pH – acid pH accelerates stabilizer migration.
- Higher concentrations of dissolved salts increase the migration rate.
- Temperature under service – higher temperatures increase the migration rate.
- S:V – as migration is diffusion controlled, a large S:V provides a larger transfer area.
- Retention time – retention time does not affect the rate of migration, but it impacts on the final concentration of the stabilizer, as the metal is allowed to accumulate in stagnant water.
- Solvent-cement joints may increase stabilizer migration.

In conclusion, the methodology available for identification and determination of the effects of stabilizer migration on water quality needs to consider the worst-case scenario based on the quality of water that can be found during service.

In particular, differences in stabilizer concentration can arise if the S:V used in the laboratory experiments is not related to the actual S:V of the product during its application.

Water quality conditions, such as pH, need to be considered. Most standard tests specify a slightly acidic extractant solution, and use extraction in stagnant water for a minimum of 24 hours, which covers the period of highest migration rate.

The standard methodology currently available has very similar procedures for the assessment of stabilizer migration. Methods such as those outlined in AS 4020 and ISO 8795 are considered appropriate, provided consideration is given to the water quality conditions typical of the areas where the pipe is to be employed. Once again, it is suggested that the worst-case scenario be considered during evaluation.

ANSI/NSF 61 is very thorough in its requirement for health hazard evaluation, however, the extraction methodology could be simplified for time effectiveness without detriment to the hazard determination. It is recommended that the preconditioning period for the test sample in water be shortened from 14 days to 24 hours, as in ISO 8795, if pipes are subjected to flushing periods during installation or, if more rigor is required, that the test sample can be subjected to rinsing, as in AS 4020, followed by analysis of the first water sample collected after 24 hours to determine the most severe leachate rate. This would then be followed by the normal extraction procedure outline in ANSI/NSF 61.

As for water quality, two approaches can be used – either the most stringent water quality conditions (pH, TDS, chlorine, hardness) (which cause the most leaching) are used, or the water quality of the test could be tailored to the geographical area where it will be used, accounting for differences in water hardness or pH, as in Europe. ANSI/NSF 61 tends to follow the first approach in the specifications for water quality of the extractant solution.

Regarding the use of solvent-cement joints, it is recommended that evaluation of the migration of stabilizers be determined for small diameter pipe, and that consideration be given to the time required for the joint to cure before putting the pipe into service, as solvent cement appears to increase stabilizer mobility.

For tin- and organic-based stabilizers, further investigation into their toxicity may be required to determine if a maximum contaminant level is applicable (current information does not suggest these substances are harmful to humans in the concentrations detected in potable water). However, the concentration of metal stabilizers from PVC pipes detected in potable water is generally below the detection limit, hence the health risk is minimized.

Finally, the procedure of flushing the pipe under turbulent flow before use is strongly recommended to remove excess depleted stabilizer from the pipe surface. Moreover, flushing is standard good construction practice, as it removes unhygienic materials that may have been deposited in the pipe during installation.

CHAPTER 2
LITERATURE REVIEW AND REPORT – WEATHERING AND THERMAL DEGRADATION IN PVC PIPES

INTRODUCTION

Degradation of PVC can occur by either thermal or UV processes, commonly referred to as thermo-oxidative and photo-oxidative degradation. Thermo-oxidative degradation is of concern when the pipe is extruded at elevated temperatures, and when the pressure pipe continuously operates at elevated temperatures. Photo-oxidative degradation is of concern when pipe is exposed to sunlight for long periods. Although initiation of the two degradation pathways (thermo- and photo-oxidative) differs (Gardette and Lemaire 1991, Osawa and Aiba 1982), the chemical processes in both forms of degradation are accepted to be generally the same (Veronelli, Mauro, and Bresadola 1999; Billingham and Calvert 1980; Rabek, Ranby, and Skowronski 1985). In both thermo- and photo-oxidation of PVC, the formation of polyenes or conjugated dienes via the dehydrochlorination or zipper-elimination process, and the competing oxidation process in the presence of air, occur (Real and Gardette 2001a, 2001b, Qureshi, Hamid and Maadhah 1989). In outdoor applications where PVC is used, thermo- and photo-oxidative degradation processes typically occur concomitantly (Rabek, Ranby, and Skowronski 1985).

In pipe and other thick PVC profiles, degradation is controlled by oxygen diffusion and is auto-retarding, confining degradation in most cases to the surface at the micron level (Fedoseeva et al. 1981).

In outdoor exposure of PVC, photo-oxidation occurs via absorption of UV radiation typically in the range of 290–400 nm (Carr 1978, Swasey 1980, Ragab and Alawi 2001). Pure PVC is not expected to absorb light of wavelengths greater than 190–200 nm, and thus PVC exposure to visible light should not cause any degradation, unless chromophores and other structural such as hydroperoxides and carbonyls are already present. Both processing impurities and defects created during polymerization processes allow for UV absorption above 200 nm. This absorbed UV energy facilitates the first stage of photodegradation – the dehydrochlorination process, which, as the name implies, yields hydrochloric acid (HCl). Initiation occurs by chain scission at labile chlorine atom sites, such as allylic and tertiary chloride, and at oxovinylene groups in the chain (which contain three types of active centers CO, C=C and CCH_3), all of which are caused by chain termination and chain transfer reactions during polymerization, and during processing at elevated temperatures. In the absence of oxygen, large polyene sequences are created, but in oxygen the sequences are hindered. Oxidative degradation, in general, accelerates dehydrochlorination in PVC by generating radicals (Rabek, Ranby, and Skowronski 1985). Reactions in the presence of oxygen are complicated; they lead to free radical chain reactions, chains scission, crosslinking and, of course, oxidation (Real and Gardette 2001a, Gumargalieva et al. 1996, Prinet–Anton et al. 1998, Gardette and Lemaire 1991). Of the many byproducts and intermediates formed, colored and easily oxidizable conjugated polyene structures and HCl by dehydrochlorination are generated. Once an olefinic or diene group is introduced on the PVC chain, dehydrochlorination is further activated by the neighboring allylic chlorine atom, as well as generated HCl acid, to form longer polyene sequences (Amer and Shapiro 1980; Verdu 1978). The presence of HCl has an autocatalytic effect on the degradation of PVC, and is temperature dependant (Amer and Shapiro 1980, Verdu 1978). The absorption spectrum of PVC is changed upon polyene formation, and the number and size distribution of sequences is responsible for the coloration in PVC (Gardette, Gaumet and Philipart 1993; Bacaloglu and Fisch 1995; Gardette and Lemaire 1991; Thomas 1993). Progressive discoloration occurs due to the formation of polyenes in PVC and for a non-pigmented resin in the following order: colorless to light yellow, to yellow to yellow–orange, to orange–red, to red, to brown, and to black, and as degradation increases the mechanical properties also subsequently deteriorate (Thomas 1993).

Thermo-oxidative degradation of PVC may occur in the initial polymerization process of PVC (Naqvi 1985) where, due to excessive temperature and in the absence of appropriate stabilizers, radical reactions lead to sites of unsaturation and oxidized polymer chains, creating inherent sites for initiation of both thermo- and photo-oxidative degradation of the finished PVC product. During compounding of PVC resin in pipe applications, the presence of inherent defects introduced during polymerization, and radical chains created by scission of labile C-Cl bonds, facilitates polyene formation and the general degradation processes, as described previously for photodegradation (Naqvi 1985). Initiating chain scission reactions in the thermal degradation process are possible due to the high heat energy associated with elevated processing temperatures and shear rates, especially in the absence of efficient heat stabilizers.

In outdoor, above-ground applications (and in some cases underground, e.g., Saudi Arabia (Naqvi 1985, Qureshi et al. 1989)), both thermo- and photo-oxidative degradation occur due to direct weathering and heat exposure. The two processes compliment and mutually catalyze degradation.

MECHANISMS OF PVC CHEMICAL DEGRADATION

There is general agreement that PVC degradation occurs via a process called "zipper elimination" (Minsker, Kolesov, and Zaikov 1988; Yassin, Sabaa, and Mohamed 1985). This process involves the initial slow elimination of HCl, and the formation of isolated double bonds, followed by a fast dehydrochlorination activated by allylic groups formed to create conjugated polyenes consisting of between 1 and 25 double bonds. The mechanisms of the individual steps in this process are controversial and (based on experimental evidence) free radical, ionic, and concerted molecular reactions are proposed (Bacaloglu and Fisch 1995). It is known from experimental trials that the rate constants of PVC degradation are linear and dependant on the concentration of HCl (Bacaloglu and Fisch 1995).

Free Radical Mechanism

Radical chain mechanisms involving initiation from catalyst residues or oxidation, which results in H abstraction from a methylene in the polymer, results in the formation of isolated double bonds, and thus is considered feasible for the formation of polyenes (Starnes 1981). A more likely radical mechanism involves chain scission of a labile C-Cl bond leading to H abstraction from the adjacent carbon atom, creating an organic radical with conjugation reoccurring by chain scission (Figure 2.1). The process may be stopped by radical combination between different molecules (crosslinking), or a Cl or H atom from another molecule (chain transfer), thus restricting the length of the polyenes formed. This mechanism is postulated to occur at high temperatures, i.e., at 200°C (392°F), with the concentration of free radicals increased with increases in temperature (Hay 1970).

Ionic Mechanism

A proposed ionic mechanism involves the ionization of a Cl atom, and the rapid elimination of a proton. The formation of an allylic Cl then activates ionization to cause a rapid chain reaction (Figure 2.2). The ionic mechanism involves a polar transition state, rather than free Cl and carbonium ions. Bacaloglu and Fisch (1995) state that the ionic mechanism proposed does not explain the restricted length of polyene sequences, and termination reactions of an ionic mechanism have not been proven. Using thermogravimetric/Fourier transform infra-red (TGA/FTIR) techniques to study the degradation of PVC, Pielichowski and Hamerton (1998) proposed that an ionic mechanism accounts for the restriction of the polyene sequence length. They state that the formation of a large polenyl cation concentrates the positive charge in the center of the PVC molecule, decreasing the likelihood of a chloride ion migrating to the sequence end and abstracting a proton from a methylene group. In the case of a stabilized resin, the action of intermolecular forces between the cation and the stabilizer hinders chloride ion motion (Pielichowski and Hamerton 1998).

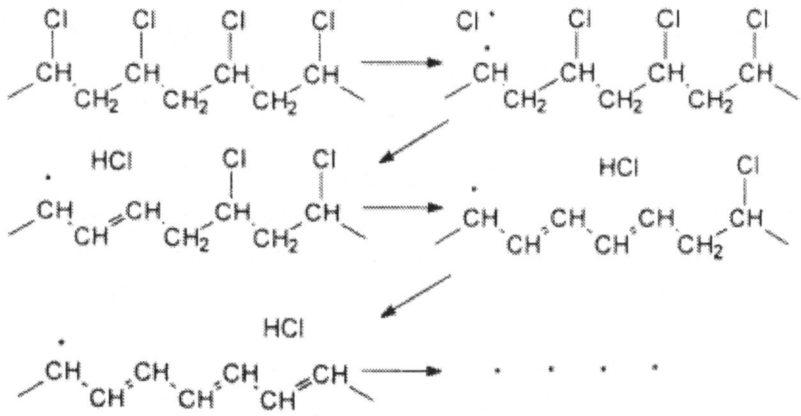

Figure 2.1 PVC degradation via a free radical mechanism (Data from Bacaloglu and Fisch 1995)

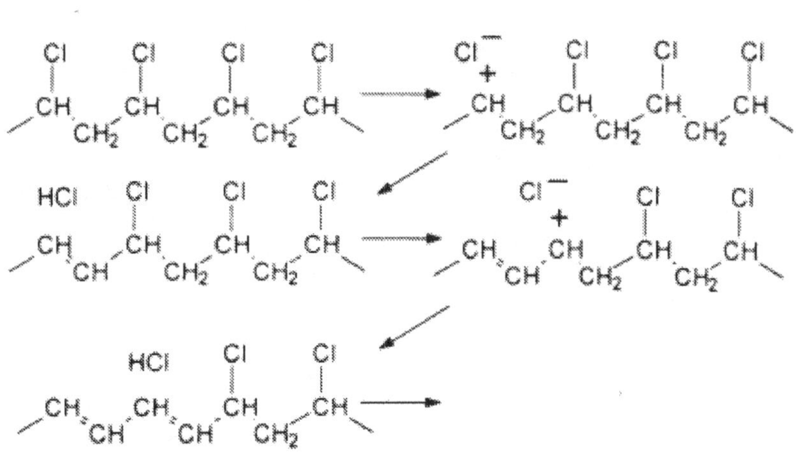

Figure 2.2 PVC degradation via an ionic mechanism (Data from Bacaloglu and Fisch 1995)

Concreted Molecular Mechanism

Another mechanism is one which involves non-ionic molecular reactions, where a slow initiation forms random double bonds on the polymer by a free radical or 1,2 unimolecular elimination of HCl. The next step involves a six-center transition state where HCl is eliminated via a cis-γ-alkyl-allyl chlorine, forming a conjugated diene, after which an HCl catalyzed proton rearrangement occurs to form another cis-γ-alkyl-allyl chlorine. The process is terminated when a proton rearrangement occurs to form a trans alkyl-allyl chlorine (Figure 2.3). The process is said to continue as long as the polymer possesses syndiotactic conformations and lacks other structural defects. This mechanism supports the autocatalytic role of HCl in the degradation of PVC (Amer and Shapiro 1980).

By use of molecular orbital calculations, Bacaloglu and Fisch (1995) describe the conditions under which the ionic, radical, and molecular mechanisms apply. They suggest that several different mechanisms apply as a function of the degradation conditions. In low-temperature inert atmospheres, they propose a succession of molecular concerted reactions to be responsible for PVC degradation. In the presence of strong Lewis acids, an ionic mechanism occurs, with the formation of carbocations via C-Cl heterolysis. The presence of such acids causes extremely rapid degradation of the PVC. The Lewis acids described may be formed depending on the types of stabilizers used. Finally, in conditions of elevated temperatures, such as those that occur during processing, or during irradiation and weathering, radical mechanisms are proposed to be predominant. This also induces fast elimination of HCl and accelerates PVC degradation. Benavides et al. (1997) suggest that at elevated temperatures (180°C/356°F), the efficiency of stabilization depends on the complexes formed between the metal stearates added, and the interaction with PVC molecules, indicating an ionic mechanism.

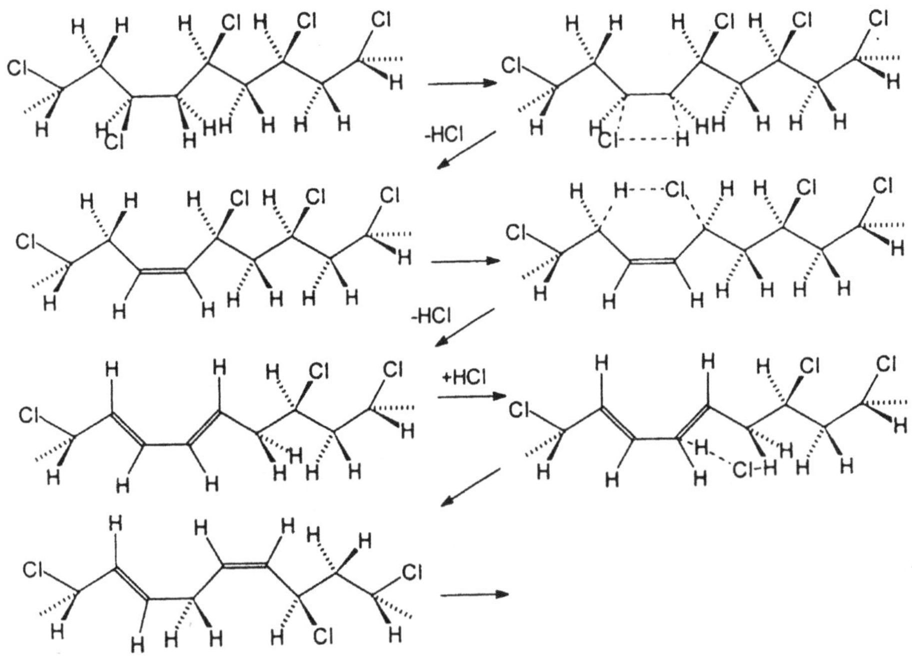

Figure 2.3 PVC degradation via molecular concerted reactions (Data from Bacaloglu and Fisch 1995)

STABILIZATION AGAINST THERMO– AND PHOTO–OXIDATIVE DEGRADATION IN PVC PIPES

Several possibilities exist to combat chemical degradation in the form of thermo- and photo-oxidative degradation in pipe applications. These include chemical stabilization in the form of heat and weathering stabilizers, additives and inorganic fillers, polymer blending, polymer modification by grafting, and even painting of outdoor pipes with UV-protective coatings (acrylic paints are known to be extremely effective). These options are viable for some applications and not viable for others when considering the cost/performance ratio, due to the required stabilizer loadings, changes to operating conditions, and maintenance costs.

Examples of applications that meet the cost/performance requirements are vinyl sidings, above-ground irrigation pipes, above-ground mine slurry pipes, fencing, and decking.

For the purposes of this paper, we will concentrate mainly on the chemical stabilization of PVC pipes intended for burial. For PVC pipe resins, stabilization in the form of antioxidants and phosphites begins in the polymerization process to reduce the number of inherent polymer defects created (Qureshi et al. 1989). The critical stage of polymer stabilization to prevent thermo-oxidative degradation is during polymer compounding, and in the extrusion process. However, even if efficient heat stabilizers are included in a pipe formulation, weathering durability cannot be achieved if degradation occurs during processing and extrusion of the pipe due to high temperatures (Mur 1995). Traditionally, metal-based stabilizers, such as Pb stearates, and other stearates based on Cd, Sn, and Ba metals, have been used as heat stabilizers during processing to ensure the service life of PVC pipes (Cuilleret 1998; Berger 1989). In many countries Pb based stabilizers are quickly being replaced due to concerns regarding the effect on water quality, as discussed in Chapter 1. More recently, combinations including alkaline earths such as Ca and covalent Zn metals have replaced traditional metals such as Pb. Although there is concern about the use of metals in general, these metals are viewed as being more environmentally friendly. The approach in PVC pipe stabilization is rapidly heading towards Ca/Zn formulations over the traditional heavier metals such as Pb and Cd, and even more so towards metal-free stabilizers such as OBSs offered by vinyl-additive manufacturers such as Crompton (Hopfmann 2002). At the European Stabilisers Producers Association (ESPA 2002), the following was presented as a voluntary commitment from the European PVC Industry:

- To completely phase out cadmium stabilizers (already achieved in the EU, Norway and Switzerland by March 2001).
- To decrease the use of lead by the following targets – 15% by 2005, 50% by 2010 and 100% by 2015 (based on 2000 consumption levels).

However, it was also stated that Sn stabilizers are not only safe when appropriately handled by industrial users, but specifically octyl and methyl tin stabilizers are broadly used for food contact applications. A distinction between TBT biocide compounds and mono- and dialkyl tins used in polymer additives must be made (Rosenthal 2002). Zn is not considered a problem by the ESPA, however contamination of soil and sediment areas by Zn is still considered to be an issue, and a risk assessment has been extended to 2003. Zn stabilizers based on plastics and Food Contact Directive 90/128/EEC, fall into two marketing categories: organics and metals without Zn, and organics and Zn metals (Rosenthal 2002).

Chemical Stabilization of PVC Pipe

Effective chemical stabilization of PVC will cause a reaction at sites of structural abnormality, preventing the reactive site from contributing to dehydrochlorination. Reactive groups such as C=O and C=C, are either destroyed, or labile sites are replaced with a more stable group. This is possible by the use of metal stearates, epoxy compounds or maleic anhydride, organic phosphites, or dienophiles. Stabilization compounds, which prevent dehydrochlorination and other degradation effects such as crosslinking, provide increased stabilization, since they also improve the processing tolerances of the PVC resin. Since the chemical reactions in thermo- and photo-oxidative degradation reactions are similar, stabilizers used in PVC pipe processing to prevent thermo-oxidative degradation (such as metal-based stearates) are also used in stabilization against the effects of weathering. As will be discussed later, the two degradation processes require independent stabilizing additives, such as anti-oxidants and phosphites for high temperature and shear pressure conditions during processing, and UV absorbers and fillers for light stabilization. The formation of polyenes in PVC by thermo- and photo-oxidative degradation creates surface discoloration, which at later stages is associated with deterioration of mechanical properties. Thus, stabilization success is often measured by the ability of the stabilizer to impart color retention.

Lead-based Stabilizers

Lead-based stabilizers undergo complex reactions that result in the substitution of the labile chlorine atoms in the polymer chain, eliminating initiation sites. Additionally, the stabilizer is able to react with HCl to form lead chloride, which is a metal salt that does not destabilize PVC, as some other metal salts do (e.g., Ca and Zn). As a result, the formulation does not require the addition of co-stabilizers (Gächter and Müller 1985).

The most common Pb-based stabilizers used in PVC pipes are lead salts or soaps such as tri-/tetra-basic lead sulfate (heat stabilizers $3PbO.PbSO_4.H_2O$, and $16PbO.4PbSO_4.H_2O$ respectively), di-basic lead phosphite (heat and UV stabilizer $4PbO.2PbHPO_3.H_2O$), normal lead stearate ($Pb(C_{17}H_{35}COO)_2$), and di-basic lead stearate (stabilizer and lubricant $2PbO.Pb(C_{17}H_{35}COO)_2$). Mixtures of tri-basic lead sulfate and di-basic lead phosphite also provide excellent weathering properties.

Lead stabilizers provide good heat and light stability, good electric properties, low water absorption, a wide processing range, good short- and long-term mechanical properties, and have a low cost to performance ratio. Some lead salts, however, are prone to sulfur staining and are not suitable for clear applications (Schiller and Fischer 2004).

Despite such advantages, lead stabilizers are being phased out of the pipe industry because of political and environmental concerns. The greatest risk of lead exposure through PVC is by inhalation of stabilizer dust during PVC manufacture, however this has been minimized in industry by providing lead stabilizers in dust-free forms (e.g., pellets, flakes, or granules) that are already mixed with stearates and lubricants. The concentration of lead stabilizers in PVC pipes is small (between 0.5% and 2.5%), and most of the lead in pipes is bound to the PVC matrix. Furthermore, the majority of the inorganic lead salts used; have poor solubility in water, and only become labile if the pipe is incinerated.

Tin-based Stabilizers

Although tin stabilizers are not considered to be as environmentally toxic as lead-based stabilizers, they are considered to be ecotoxic and not as desirable as Ca/Zn stabilizer combinations (Balkose et al. 2001). There are two types of tin stabilizers – maleates and mercaptides. Tin mercaptides are the optimum for processing under high temperature and shear conditions, such as those in pipe and fittings manufacturing (Morel, Cuilleret, and Girois 1999). However, they do not protect against photodegradation as well as maleates. Thus, both of these stabilizers are used successfully in combination for PVC pipe applications. All heat stabilizers neutralize HCl, inhibiting its dehydrochlorination autoxidation. Tin stabilizers have an anionic group that is substituted with labile chlorines in PVC. As the anionic groups are less reactive, the unzipping reaction is retarded. Organotin compounds are also anti-oxidants that suppress the formation of hydroperoxides. Common tin-based stabilizers are usually in the form of alkylated tin maleates, such as dialkyl tin maleate. A new generation of solid tin maleate stabilizers for outdoor stabilization of rigid PVC is available in the window industry (Morel, Cuilleret, and Girois 1999). The specific structure of the stabilizers allows for excellent weathering properties and ease of processing (as opposed to liquid systems, which were sticky and volatile due to low molecular weights). Contrary to mercaptides, tin maleates stabilize PVC dienes via Diels–Alder reactions. Although the mechanism is not completely understood, tin maleates have shown excellent color retention in comparison to Pb and Ca/Zn systems, in both white and pigmented window profiles. Initial gloss and impact properties are also retained at a stabilizer addition of 0.2–0.5%.

Ca/Zn Stabilizers and Organic Co-stabilization

A feasible replacement for lead stabilizers is a combination of Ca/Zn stearates in conjunction with organic co-stabilizers such as polyols, and diketones (Thomas 1993). Several manufacturers in Australia have successfully optimized the Ca/Zn formulation and use such systems in rigid pressure pipes. These stabilizers impart better heat and light stability with UV absorbers and anti-oxidants, and must be used in conjunction with synergistic co-stabilizers such as beta diketones, hydrotalcites, and zeolites. The alkali earth metals (Ca) impart long-term heat stability but not initial color retention. When mixing both Zn-based covalent stearates (which impart good initial color stability) and alkali earth metals, both initial and long-term color retention are achieved.

The Frye Horst mechanism explains the stabilizing action of covalent stearates (Cd and Zn) by esterifying displacement of labile Cl atoms (i.e., carboxylation of labile sites (Thomas 1993)). However, the metal chlorides produced cause further degradation by alkylation of polymer chains with polyene sequences leading to crosslinking, further HCl evolution, and rapid property deterioration. The presence of alkaline earth carboxylates causes ester exchange with the covalent Lewis acids formed, leading to regeneration of the covalent carboxylates and alkaline earth chlorides which do not promote dehydrochlorination. The ratio of Ca/Zn stearates should be chosen to impart the preferable color stability, i.e., initial or long-term. A ratio of 2:3 by weight imparts initial color retention, while a ratio of 4:1 gives better color retention at a later stage of degradation, with slight initial discoloration. Since calcium stearate decreases gelation time and therefore heat stability, high Ca stearates are often proposed for optimum processing conditions. However, they impart poor initial color retention. For pipe processing, a balanced ratio is required, and this ratio must consider the complete processing package, including other stabilizers, lubricants, and pigments. For rigid pressure pipe, the ratio of Ca/Zn should be between 4:1 and 1:4 by weight (Thomas 1993).

When choosing stabilizers for processing of PVC-U pipe, consideration of the brittle failures that may occur in the field, due to processing conditions, is mandatory (Thomas 1993). Critically, the gelation process needs to be controlled. As outlined in Chapter 3, a homogeneous melt requires high temperature and high shear rates without initiating degradation. When using less efficient Ca/Zn stabilizers compared with traditional Pb-based stabilizers, optimizing processing conditions is more difficult, and thus the appropriate lubrication package must be carefully chosen. Zinc stearates are used as external lubricants in processing, whereas Ca stearates are internal and decrease fusion time, creating highly viscous melts. The use of Ca stearate can interact with waxes and cause plate out of other additives (Lindner and Worschech 1986). Thus, changes in the Ca/Zn stearate system not only affect the thermal stability of the polymer during processing, and both thermal and light stability in service, but also the processing conditions and the pipe end product as a result (Aylett 1990).

Several patents exist describing the use of organic co-stabilizers or secondary stabilizers with metal-based stabilizers for stabilization of PVC resins (Scrima and Vessa 1995, Beekman 1996, Baker and Grossman 1996). Secondary stabilizers increase the efficiency of the Ca/Zn metal stearates or soap combinations. They include polyols (e.g., glyceryl monostearate) that are added in partially esterified form to prevent moisture. They improve initial color stability and severe discoloration catalyzed by the presence of $ZnCl_2$. The mechanism of action for polyols is unknown, but those that have been proposed suggest that the secondary stabilizers react directly with HCl, removing these catalysts, and thus the resultant dehydrochlorination, from the system. This allows for higher zinc soap content in the system, and thus the prevention of initial discoloration. Most polyols lower the viscosity of the polymer mixtures. Combinations of pentaerythritols and trimethylol propanes were quite successful. Beta-diketones are thought to act in a similar manner, and also by substituting themselves for labile Cl atoms in the polymer chain (Michel et al. 1981). These react quicker in the presence of Zn chlorides. Systems incorporating secondary stabilizers (which act by substitution) require further secondary stabilizers that scavenge HCl as Zn chloride and HCl accumulate. Secondary stabilizer polyols such as zinc mercapto acids (zinc bis-thioglycolates, zinc bis-mercaptoproprionates, and mercaptosuccinates of glycols) eliminate the phenomenon of zinc burn during processing (Beekman 1996). Terminal epoxides (glycidyl and b-methylglycidyl esters) are also added as secondary stabilizers in the liquid (soy bean oil), or as a solid coated on hydrotalcites or on PVC (masterbatch) (Drewes et al. 1996). Carbazoles hinder the evolution of HCl by an aromatic alkylation reaction (Pielichowski and Hamerton 1988). Pyrrolidine-2,4-diones (P24D) have been shown to be the most effective co-stabilizers as compared to other b-diketones typically used (including dibenzoylmethane and stearoylbenzoylmethane) for mixed metal stabilizers replacing lead. The additive P24D is non-toxic and is included at low levels. It is thought to disassociate from the metal co-stabilizer complexes, increasing the efficiency of the stabilizing system (Burley 1997).

Organic-based Stabilizer Systems

OBSs are the more recent (mid-1990s to today) development of completely metal-free organic stabilizer systems for PVC pipe formulations (Pfaender et al. 1998, Hopfmann 2002).

These are solid multi-component blends based on patented organic heterocycle molecules (Figure 2.4), termed as pyrimidinediones or uracils, with highly nucleophilic sites, in combination with synergistic co-stabilizers. They have been shown to provide good resistance to discoloration in PVC, and to maintain tensile and mechanical properties.

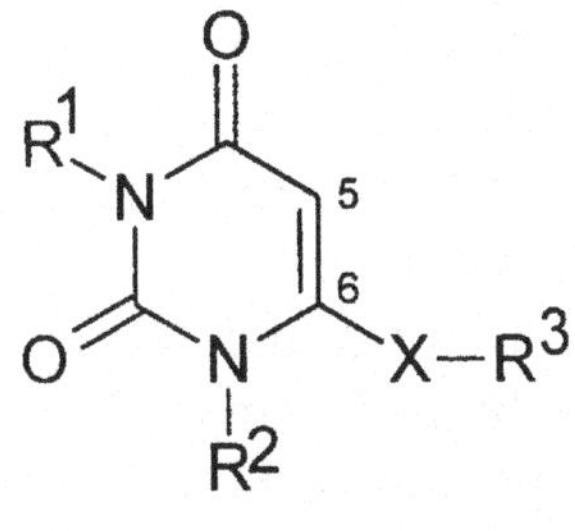

Figure 2.4 Chemical structure of organic pyrimidinedione compound, where R^1 and R^2 are alkyl groups, XR^3 is an electron-donating group, and R^3 is an alkyl, phenyl or hydrogen substituent (Data from Hopfmann 2002)

Co-stabilizers in these systems act as HCl scavengers, and are usually inorganics or, in some cases, alkali earth metal stearates. The German additive manufacturer, Crompton Vinyl Additives GmbH, claims to have developed an OBS system with a good cost/performance ratio (Hopfmann 2002). OBS systems have shown excellent processing behavior in the PVC window profile industry, and also resistance to thermal and light degradation by accepted evaluation methods such at the Congo red test. Initial mechanical properties of pressure pipe, such as impact strength, fracture toughness, Vicat softening temperature, and burst pressure tests have also been shown to be acceptable by European Standards (Table 2.1) (Hopfmann 2002).

Table 2.1
A comparison of mechanical properties of 110-mm pipe, stabilized with OBS and lead-based systems (Data from Pfaendner et al. 1998)

Properties	Standard	Lead stabilized	OBS
Fracture toughness	BS 305	4.2 MPa	4.2 MPa
Vicat temperature	EN 727	83–84°C	84°C
Vicat softening point	PrEN 20075	74°C	73°C
Impact strength B50; 4 kg (0°C)	ISO 11173	222 cm	213 cm

The performance of OBS systems is due to the conjugation in a heterocyclic molecule, where electron transfer within the ring allows for retardation of the dehydrochlorination reaction, preventing (or shortening) the formation of polyene sequences. Figure 2.4 shows the basic structure of an uracil-type stabilizer. The unsubstituted site in ring position 5 is beneficial when combined with the electron-donating group in position 6. The effect of the stabilizer is further enhanced when R^1 and R^2 are substituted as alkyl and phenyl groups.

The introduction of OBS systems in the PVC industry has been successful in altering the “processing window” of window profile extrusion (Crema and Aylett 2002). Research is being conducted on the influence of Ca/Zn stearate stabilizers in combination with OBSs, on the processing conditions in the pipe industry, particularly for modified PVC (PVC-M) and orientated PVC (PVC-O). However, the use of such stabilizer systems in these high-performance PVC pipes will present a greater challenge than their use in PVC-U. Ca/Zn stearate-based stabilizers have been on the Australian market for both PVC-M and PVC-O pipe for over a decade. Changes to these systems to include OBS components may change the formulation and thus the processing window of the PVC formulations. The processing window is defined as the range of operating conditions which can be achieved to obtain specified physical and aesthetic properties concurrent with adequate stability, using the available processing equipment at the lowest possible cost. PVC residence times in processing follow the order PVC-M > PVC-O > PVC-U. Thus, PVC-M places the highest load on stabilizer systems. During cooling, the product with the thickest wall will have the highest thermal inertia, which also places a high demand on thermal stabilizers. Furthermore, PVC-O places an even greater demand on stabilizers as it must be reheated for molecular orientation (Crema and Aylett 2002). Although PVC-U pipe usually has thicker walls, PVC-M places a higher demand on its stabilizers due to increased heat retention from higher processing temperatures. Studies of specific shear and melt temperatures (also internal/external lubrication and residence time (Crema and Aylett 2002)) during processing indicate the level of stability in PVC, and can be used to determine the limits of an OBS stabilizer system. Specific melt temperature ranges and shear energies (extruder machine torque) will provide reliable physical and aesthetic properties for PVC grades, and these conditions will define the processing window. By modification of Ca/Zn and OBS systems, the processing window is widened in the direction of the lower processing temperatures (195–200°C). Initial work with OBSs has shown that these systems have similar processing windows to other systems. The processing window based on Ca/Zn stabilizers is narrow when considering both mechanical properties and thermal stability. However, the current cost to achieve the wider processing window offered by OBS as compared to Ca/Zn stabilizers is relatively high (Hopfmann 2002).

Anti-oxidants, UV Absorbers and Inorganic Fillers

Anti-oxidants are included to prevent oxidative degradation in PVC-U, which may be initiated by heat, weathering, mechanical stress, or catalyst residues. Oxidation in PVC adds to the dehydrochlorination process. Anti-oxidants act as free radical scavengers, and anti-oxidants such as hindered phenols are used at low levels. Their addition must be such that anti-oxidants of this kind do not cause staining in the presence of additives such as titanium dioxide, metal oxides, or hydroxides.

During oxidative degradation, alkyl radicals further react with oxygen to accelerate the degradation process. Anti-oxidants interrupt the chain process by two means: primary anti-oxidants (such as sterically hindered phenols) trap radicals, and secondary anti-oxidants (such as phosphites) decompose reactive hydroperoxides or oxidation products into non-reactive products. They also decompose reactive hydroperoxides or oxidation products into non-reactive products. Anti-oxidants are also necessary in the polymerization process. Some highly hindered phenols are not as efficient at trapping radicals, and can therefore be used in the vinyl polymerization process with only minor effects on the polymerization process. The product of polymerization will have improved color and thermal stability. High molecular weight anti-oxidants are more effective because they are less volatile and are not lost during the drying process of PVC resin. Thus, they provide greater stability during processing and compounding of the PVC resin.

Inorganic phosphites have a synergistic effect with metal salts, but they also have a plasticizing effect and should not be used with rigid PVC. Non-toxic calcium-aluminium-hydroxy phosphites can be used as secondary stabilizers with Ca/Zn stearates, and have been stated to provide heat stability similar to that of lead phosphite. They are hydrated, but do not release water at PVC processing temperatures, and thus are better than hydrotalcites.

Other stabilization components in PVC specific to weathering include ultra-violet absorbers (UVAs), hydroxybenzophenones, hydroxybenzotriazoles, alpha-cyanoacrylates, oxanilides, and phenyl salicylates. UVAs are efficient in protecting PVC pipe since (through a proton transfer reaction) they allow radiation to be dissipated as heat at the pipe surface.

Hindered amine light stabilizer (HALS) systems for light stabilization are not completely understood, but a proposed theory suggests that nitroxy radicals terminate and deactivate alkyl radicals and peroxy radicals that participate in photodegradation (Capocci and Zappla 1986). They do not absorb UV and thus work well in low concentrations. They are synergistic with UVAs and impart excellent early color retention in PVC.

For PVC pipes and outdoor profiles, inorganic fillers such as zinc oxide, titanium dioxide (TiO_2), and carbon black have been added at low levels to impart weathering resistance (Gesenhues 2000). Inorganic fillers such as TiO_2 impart UV resistance by screening and scattering radiation at the PVC surface (Gesenhues and Hocken 2000). The surface quality of TiO_2 plays a major role in its effectiveness as a light stabilizer or as a sensitizing agent of degradation in PVC. Furthermore, different irradiation climates have an effect on the actions of this pigment. It has been shown that PVC-U formulations with a weight loading of 2% TiO_2 retain most physical and aesthetic properties after two years of outdoor exposure (Burn 1992). However, TiO_2 absorbing UV in the presence of moisture causes hydroxyl radicals to form at the TiO_2 crystal surface. Such radicals lead to photo-catalytic breakdown of the polymer in the vicinity of the crystal. This causes adhesive failure between the additive and the polymer, which in turn produces chalking at the surface and a loss of gloss properties (Burn 1992). In such cases, TiO_2 needs to be surface treated (Gesenhues 2000). Either rutile or anatase TiO_2 is the preferred form for efficient light stabilization by these inorganic fillers (Gesenhues 2000). The titanium additive screens and scatters UV light. These inorganic fillers are also preferable at low loadings due to their cost and also due to their demand for increased shear stress during extrusion. Degradation of PVC via such pigments is based on the ability of these pigments to transfer oxygen to the polymer molecule. Stabilization is based on their ability to absorb UV, and convert UV energy to heat energy that is less likely to degrade a polymer. Synergistic stabilizers are often added with oxide pigments to prevent heat energy from degrading the polymer (Gesenhues and Hocken 2000).

POLYMER MODIFICATION

The incorporation of chlorinated polyethylenes in PVC to produce PVC-M, was intended to increase the toughness and impact resistance of the material, specifically in pipe applications (Whittle, Burford and Hoffman 2000). Another example of modified PVC morphology is in the window industry, where ethylene vinyl acetates (EVAs) have been included. While lowering the elastic modulus of the PVC, the toughness is greatly enhanced. A further benefit from the inclusion of EVAs, is improved thermal and weathering stability (Menzel 1979). One such example of modified PVC is a window profile produced by HULS, under the name Vestolit BAU (HULS 1985). Graft polymerization of EVA onto the PVC-U polymer chain is performed to enhance light fastness and weathering resistance of the polymer. HULS grafts EVAs and polyacrylic esters (both having excellent UV resistance), therefore increasing the weathering resistance of the modified PVC. These grafts also allow lower processing temperatures, which in turn enhances heat stability and weathering resistance (HULS 1985). The cost to performance ratio for polymer modification in PVC pipes may be a viable option in extreme climatic conditions where pipe and fitting failures are excessive and premature.

QUANTIFYING THE EFFECTS OF DEGRADATION ON LONG–TERM PERFORMANCE OF PVC PIPES

Several analytical and mechanical methods are available for the assessment of both thermal and weathering degradation in PVC polymers, most of which are applicable to PVC pipe. These techniques include infra-red (Jouan and Gardette 1987; Butters and Marks 1978) and UV absorption spectroscopy, gas chromatography and pyrolysis, fluorescence spectroscopy (Breant and Stapfer 1994), thermal analysis by differential scanning calorimetry (DSC), dehydrochlorination, impact strength assessment, tensile strength, fracture toughness testing, and microhardness testing (Deshmukh and Lai 2000; Ragab and Alawi 2001). It must be noted, however, that the majority of these techniques are based on short-term exposures. By monitoring changes in material properties with exposure time (for both weathering and thermal degradation), a combination of these techniques can be used to identify degradation mechanisms. However, they are not adequate to allow for prediction of pipe performance properties (such as burst pressures in PVC pipe with aging or exposure time). Natural weathering or thermal aging for periods of up to five years may often produce only minimal changes in material properties in benign climates, and the use of accelerated conditions distorts the true effects of degradation on PVC property performance at ambient operating conditions.

The effects of thermal and weathering degradation of PVC pipe applications are mostly surface effects, but nevertheless they have critical consequences on both physical and mechanical properties. In a two-year study in the United States, it was found that thermal degradation and weathering significantly reduced the impact strength of PVC pipe. The modulus of elasticity and tensile strength were not affected (Uni-Bell 1981). However, a later study carried out in Australia (Burn 1992) found that maintaining these properties depended on the presence of a suitable level of stabilizer such as TiO_2.

The efficiency of stabilization additives in PVC pipes against thermal degradation may be assessed by one of the available ISO dehydrochlorination methods, or by the assessment of discoloration by gloss and color measurements. Similarly, the efficiency of incorporated UV stabilizers may also be assessed by the extent of discoloration that occurs during exposure to UV radiation. Mechanical properties such as impact strength, tensile elongation, and tensile yield stress, may also be monitored, and the extent of deterioration over time assessed. However, Tiganis, Burn and Davis (2002) have shown that the long-term impact properties of acrylonitrile–butadiene–styrene (ABS) at ambient temperatures cannot be predicted from impact tests after exposure at elevated temperatures using the Arrhenius equation, with activation energies for different stages of degradation being dissimilar. It is expected that the degradation kinetics of PVC would also prove difficult to model using a kinetic equation, which does not account for the different effects on degradation paths at elevated and ambient temperatures, nor the changes in the degradation mechanism with time.

The change in material fracture toughness property with aging or weathering may be used to characterize the long-term performance of PVC pipe (Ragab and Alawi 2001; Tiganis, Burn and Davis 2002). The fracture toughness of a pipe material reflects the effects of both external and internal stresses, and surface damage during the pipes service lifetime, and is useful for pipe performance prediction. In pipe applications, external notches are created by installation damage and extrusion defects. Weathering of these external surface flaws, and in many cases the flaws themselves, act as sites that initiate premature brittle fracture in burst tests (Ragab and Alawi 2001; Tiganis, Burn and Davis 2002). As the exposure period increases, the *J*-crack growth resistance for weathered notched PVC pipe decreases. This performance reduction is shown to occur in relatively short natural aging periods (up to two years) (Ragab and Alawi 2001). It is suggested that the ratio of the strength of a notched weathered pipe specimen to that of an unnotched specimen reflects the degradation of PVC pipe due to natural weathering. This is possibly due to a transition from semi-ductile (post-yield crack growth) to brittle fracture, which has been shown to occur under static loading in ABS (Tiganis, Burn and Davis 2002). By combining this information with results from other assessment methods (e.g., color analysis, FTIR, and dehydrochlorination), quantitification of the physical degradation process in a polymer is possible. While a similar model may be developed for PVC pipe, it would be specific to PVC grade, stabilizer and exposure type.

Since the thermal degradation of PVC at elevated temperatures is also a surface effect, and proceeds via the same degradation reactions, the effects of elevated temperature exposure on fracture toughness is expected to be similar to that of weathering. However, Seldén (1987) has shown that the large reductions observed in fracture toughness by aging at elevated temperatures, do not correspond to significant reductions in pipe lifetime when the results are extrapolated to 20°C (68°F) using the Arrhenius equation. However, it should be noted that use of the Arrhenius equation to predict the fracture toughness of PVC-U at ambient conditions, from data collected at elevated temperatures, again does not consider the changes in the kinetics of degradation.

In light of these findings, and the limitations of prediction techniques, a procedure for predicting weathering and thermal degradation effects on the performance of PVC pipe should be based on an assessment of the key properties that govern pipe service lifetime. These key properties should measure degradation levels in PVC pipe prior to failure, and will include (for example) impact strength, fracture toughness, and dehydrochlorination tendency as measured by pH. However, further development of the models to ensure accurate predictions of these long-term properties from short-term tests, or accelerated exposures is necessary, and it needs to be emphasized that these predictions will be formulation specific. Numerous citations in the literature clearly demonstrate that a clear correlation between natural weathering/heat aging of PVC to artificial methods has not been established (Tiganis, Burn and Davis 2002; Seldén 1987; Lewis 1993). This correlation needs to be determined if assessment methods based on artificial exposures or property extrapolation are to provide accurate indications of PVC pipe performance.

While the above discussion on failure mechanisms for pipes exposed to heat or UV exposure is relevant for the long-term performance of pipes under aggressive installation conditions, it should be emphasized that most buried pipes installed in the USA are not subject to prolonged exposure to UV radiation or heat, and therefore these exposure elements do not play a major role in determining the longevity of PVC pipes. It is only when pipes are inadequately stored or do not have adequate stabilizer added during processing that these degradation mechanisms come into play to reduce the expected life of PVC pipes.

CONCLUSIONS & RECOMMENDATIONS

The requirements for heat and weathering resistance of PVC pipe are extremely well documented, and any procedure for predicting weathering and thermal degradation effects on the performance of PVC pipe should be based on an assessment of the key properties that govern pipe service lifetime. These key properties should measure degradation levels in PVC pipe prior to failure, and should include (for example) impact strength, fracture toughness, and dehydrochlorination tendency as measured by pH. However, to ensure accurate predictions of these long-term properties from short-term tests, or accelerated exposures, additional experimentation needs to be carried out and it needs to be emphasized that these predictions will be formulation specific. Examination of the literature shows that a clear correlation between natural weathering/heat aging of PVC to artificial methods has not been established. If accelerated methods are to be used, then this correlation needs to be determined if artificial exposures or property extrapolation is to provide accurate indications of PVC pipe performance.

CHAPTER 3
LITERATURE REVIEW AND REPORT – FRACTURE IN PVC PIPES

INTRODUCTION

The adoption of PVC pressure pipes has been attributed to their relatively good field performance in comparison to older metallic pipes. For example, a study in Canada (Rajani, McDonald and Felio 1995) indicated that PVC pipes have 13.5 times fewer line breaks than traditional materials, including ductile iron. In another example, Kirby (1981) compared the failure rates of PVC and cast iron pipes in the UK 18 years after PVC was first used for water and sewage transportation. In comparison to cast iron pipes, PVC exhibited no increase in failure rate as time progressed. In fact, a decreasing failure rate was actually reported with the passing of time, which can be attributed to the bathtub effect, normally where failures associated with installation occur quickly, followed by the expected failure rate associated with pipe performance. In particular, a significant improvement was found in the performance of PVC pipes after the first UK Code of Practice was introduced in 1973, addressing transport, handling, and installation of plastics pipes (BSI 1973).

Bjorklund and Janson (1981) and Bjorklund (1992) investigated the performance of PVC in water distribution systems in Sweden. As with the UK, PVC pipes manufactured after 1973 exhibited improved performance. However, in this case, the improvement was attributed to the use of PVC materials with higher gelation levels rather than better installation practices. A number of pipes that were manufactured before 1973 and which had failed in service, exhibited poor resistance to dichloromethane attack, indicating poor gelation during pipe manufacture (Bjorklund and Janson 1981, Bjorklund 1992).

In Australia, PVC pressure pipes installed in the early 1970s in a large rural water supply scheme have been exhumed and examined (Stahmer and Whittle 2001). The pipes were considered to represent Australian quality at that time. Over a service life approaching 30 years, no failures occurred apart from those attributed to third-party damage or poorly manufactured solvent-cement joints. Similar studies have been conducted in the USA (Eckstein, 1988), the Netherlands (Gons 1995), and Canada (Seargeant 1999).

However, PVC pipe failures do occur both in Australia and North America. Most failures (but not all) can be attributed to poor installation, excessive operating conditions, or third-party damage, while a few can be attributed to poorly manufactured pipes. These failures occur in different forms – in some cases, pipelines have experienced isolated bursts along significant lengths, others sustain multiple failures along relatively short pipe lengths, etc.

Almost invariably, PVC pipe failures in the field are brittle, with cracks initiating from stress concentrations in the pipe wall (Marshall and Williams 1974). As PVC pipe use increases, the economic and social consequence of these premature fracture failures also increases.

In laboratory pressure tests on PVC pipes for periods up to 10,000 hours, it is usual to observe ductile failures in which the pipe balloons and then eventually ruptures. As stated by Marshall, Brogden, and Shepherd (1998), the time dependence of this failure mode is attributed to creep deformation, yielding, and drawing of material in the pipe wall, and eventual ductile tearing. In practice, however, field failures are invariably brittle in nature (Whittle 2002). Therefore, while the yield strength of PVC is important in determining the stress-bearing capacity of a pipe, tests are also needed to ensure that a change in failure mode will not produce premature brittle fracture in the field. This chapter examines the fracture mechanisms in PVC pipes and outlines methods used to analyze failures.

FRACTURE IN PVC-U PIPES

Figure 3.1 shows a fracture failure in a PVC-U pipe. As shown, the fracture has initiated and propagated in excess of 1 m along the length of the pipe. Brittle fracture is also suggested by the removal of a section of the pipe wall, which indicates initiation of multiple cracks from the main fracture path. Examination of the fracture surface reinforces this suggestion, with no evidence of stress whitening associated with ductile deformation (Figure 3.2).

Figure 3.1 Fracture failure in a PVC-U pipe

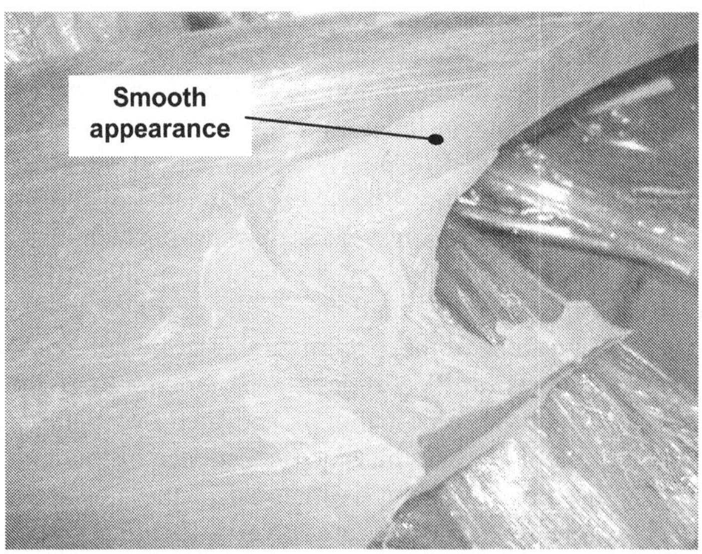

Figure 3.2 Fracture surface from brittle failure of PVC-U pipe

As shown Figure 3.2, the relatively smooth appearance of the fracture surface indicates a fast brittle fracture in the absence of ductile deformation. Examining fracture surfaces to identify failure mechanisms is an essential part of failure analysis for PVC-U pipes (Broutman, Duvall, and So 1989).

Processing and its Effect on PVC-U Pipe

Prior to pipe extrusion, PVC resin is described as comprising irregularly shaped porous "grains", approximately 100 µm in diameter (Hattori, Tanaka, and Matsuo 1972). In turn, these grains are made up of roughly rounded particles approximately 0.5–1.5 µm in diameter. These "primary" particles are bonded together within grains in a three-dimensional sponge-like lattice (Hattori, Tanaka, and Matsuo 1972). Below this size scale, primary particles are thought to consist of "domains" and "microdomains", comprised of aggregated polymer chains (Burgess 1982).

During pipe extrusion, this resin structure is broken down and re-formed, with associated influences on the properties of the finished pipe. With this in mind, there are a number of considerations for the pipe producer:

- PVC is inherently thermally unstable. Processing temperatures are invariably higher that the normal resin decomposition temperature.
- While PVC resin is largely amorphous, it does contain a degree of crystallinity that exhibits a broad melting range. Melting and re-forming this crystalline structure is essential in developing a homogeneous structure with high strength (Whittle 2002).
- The particulate structure of the resin can persist through extrusion and remain in the finished pipe, diminishing strength, and influencing failure mechanisms.

Having broken down the initial resin structure, fusion or gelation into the pipe microstructure is extremely important. Different mechanisms have been proposed for the gelation process (Menges and Berndtsen 1976, 1977), but the most recent (and widely accepted) is described below (Benjamin 1980, Parey and Menges 1981).

In the first phase of this gelation process, the molecular network remains within the boundaries of the primary particles created during resin manufacture. In the subsequent phase of the process, the application of sufficient energy (by temperature and shear during extrusion) breaks down the primary particles. The crystalline material in the PVC melts and, with its boundaries removed, the primary particle structure is converted to a continuous network of entangled polymer chains. On cooling, crystalline nuclei reform and, if molecular chains are sufficiently long, tie molecules are formed between individual crystallites, strengthening the network . The molecules will also pass through the boundary regions of the original PVC resin. The gelation process is therefore defined as the process of removing the boundaries between various particles within the resin, and the formation of a continuous mass with a homogenous molecular structure.

Irrespective of its precise mechanism, the processing conditions required for gelation are twofold. Firstly there must be sufficient shear applied during extrusion to break open resin grains. Secondly, there must be sufficient heat applied to melt at least some of the crystallites within primary particles and permit molecular diffusion (Whittle 2002).

With this in mind, the importance of gelation in the production of PVC pipes has been widely studied. Previous research has indicated that the change in failure mechanism in PVC pipe (from ductile to brittle) is related to the extent of pipe gelation (Benjamin 1980). The poorer the gelation, the sooner brittle failures occur under hydrostatic pressure, resulting in steeper regression lines describing hydrostatic pressure/time test data.

An additional aspect of processing to be considered is the influence of residual stress. Non-uniform cooling (external surface only) during pipe manufacture sets up a residual stress profile through the pipe wall thickness (Greenshields, Davis, and Leevers 1997). A region of tensile stress is created at the pipe bore, which is balanced by a region of compressive stress at the outer surface. Increasing the temperature of the cooling water first encountered by the pipe on exiting the extruder, gives a more even stress distribution through the pipe wall due to the lower cooling rates, which therefore reduces the tensile residual stress that is set up at the pipe inner wall. This substantially increases the time to stress rupture failure under hydrostatic pressure. For example, an extruded pipe is expected to withstand a hoop stress of 35 MPa (= 5076.32 psi) for not less than 100 hours (Whittle 2002). However, if residual stress can be reduced, a hoop stress of around 36 MPa (= 5221.36 psi) can be withstood for longer than 800 hours (Whittle 2002).

Analysis Methods for PVC-U Fracture

The brittle nature of fracture in PVC-U means that linear elastic fracture mechanics (LEFM) theory can be applied for analysis and prediction of failure events (Williams 1978). Qualitatively, the fracture process can be envisaged as a balance between an applied SIF and the plane-strain fracture toughness of pipe material. In general, the failure process is split into three stages:

- An incubation period between load application and crack initiation.
- A period of slow crack growth.
- Fast, brittle fracture when the applied SIF exceeds the material fracture toughness.

The applied SIF is determined by the in-service loading conditions and the "condition" of the pipe. Service loading can be split into contributions from operating pressure, external loading, and residual stress. The condition of a pipe is characterized by the size of flaws or defects in the pipe wall. Such defects could originate from external surface scratches and melt inclusions. For a defect of radial depth, a, in the pipe wall, an applied SIF, K_I, can be defined which relates the remotely applied stresses to the stresses at the defect tip:

$$K_I = Y\sigma\sqrt{\pi a} \tag{3.1}$$

where Y is a geometric correction factor to account for the influence of pipe size and loading type on K_I. Previous researchers have proposed that the incubation period before crack initiation ends occurs when the defect or crack opening displacement (COD) attains a critical value (Williams 1978, Marshall 1982). This critical COD, δ^*, is written in terms of the applied SIF as:

$$\delta^* = \frac{K_I^2}{E\sigma_Y} \tag{3.2}$$

where E is the Young's modulus, and σ_Y is yield stress.

If visco-elastic behavior is represented by the time-dependant Young's modulus $E = E_0 t^{-p}$ and $\sigma_y = \sigma_0 t^{-q}$, then the incubation time, t_i, can be written as:

$$t_i = \left[\frac{E_0 \sigma_0 \delta^*}{K_I^2} \right]^{\frac{1}{p+q}} \tag{3.3}$$

After the incubation period ends and crack initiation occurs, subsequent crack growth can be related to the applied SIF in the form:

$$\frac{da}{dt} = DK_I^m \tag{3.4}$$

where D and m are empirical constants, which can be determined from laboratory fracture tests (Davis, Burn and Whittle 2001).

Sometime during this growth period, the crack will attain a critical size and fast brittle fracture will occur. The criterion for the stress intensity at which this fast failure occurs can be expressed as:

$$K_{ITOT} = K_{IC} \tag{3.5}$$

where K_{IC} is the material fracture toughness that can be measured using a range of different test methods (ISO/DIS 1999).

Since SIFs are additive, a *total* SIF to predict this fast fracture for a pipe in service can be defined as:

$$K_{ITOT} = K_{IP} + K_{ID} + K_{IR} \tag{3.6}$$

where K_{IP} is the SIF due to internal pressure, K_{ID} is the SIF due to diametrical deflection, and K_{IR} is attributed to residual stresses in the pipe wall. In each case, the individual SIFs are described in terms of remote stresses and geometry factors of the type shown in Equation 3.1 (Lu, Davis and Burn 2003).

A rigorous analysis determines the time associated with each stages of the failure process (i.e. incubation followed by crack growth), and sums them to give the total time to failure (Lu, Davis and Burn 2003). However, an alternative approach is to simply use results from fracture toughness testing under a range of test loads (Davis, Burn and Whittle 2001; Burn 1994; Marshall, Brogden and Shepherd 1998). This can provide an empirical relationship between fracture toughness and time in the form (Lu, Davis and Burn 2003):

$$K_{IC} = At^B \tag{3.7}$$

where t is time in hours, and A and B are constants. Having established such a relationship, the actual total SIF applied for a pipe in service can be obtained from Equation 3.6, and used in Equation 3.7 to determine a time to failure.

FRACTURE IN PVC-M PIPES

While well-processed PVC-U pipes have high yield strength and modulus values, notch sensitivity (and the tendency to fail in a brittle manner in service), has restricted its usage in some cases. Considerable attention has been directed at enhancing the toughness of PVC by incorporating additives or "impact modifiers" in Australia and Europe. Toughness can be envisaged as the resistance to fracture by absorption and dissipation of energy. Hence, impact modifiers attempt to change the failure mode of PVC pipes by absorbing an increased amount of energy upon dynamic loading (Whittle 2002). While the literature has concentrated on the effect these modifiers have on impact resistance, for buried pressure pipes, their influence on crack initiation and slow crack growth is more important (Whittle 2002).

Additives for impact enhancement fall into two categories:

- Non-predefined elastomer modifiers which are uncrosslinked elastomers such as EVA and chlorinated polyethylene (CPE). These materials are added in powder form and develop their phase morphology during pipe processing. While there has been opposing views on the phase morphology of PVC-CPE blends, a more recent study in Australia (Whittle 2002) has shown that they primarily consist of discrete particles of CPE dispersed in a continuous PVC phase.
- Predefined elastomer modifiers are core-shell structures usually with a crosslinked elastomer core and a hard shell composed of one or more monomers to ensure compatibility with the parent PVC. The particle size of the modifier is established during the manufacture of the modifier itself, and does not change during pipe processing (Whittle 2002). Common examples are methacrylate–butadiene–styrene and methacrylate–acrylonitrile–butadiene–styrene.

Polymeric impact modifiers all have glass transition temperatures below the operating temperature of the finished pipe, which means they have a relatively low modulus and tensile strength compared to the surrounding PVC matrix. Consequently, the modulus and strength of a PVC-M pipe is reduced compared to a PVC-U pipe. For pressure pipes, CPE modifiers are widely used in practice, and achieve an optimum balance between increased toughness at the expense of reduced strength and modulus.

Comparative examination of failure surfaces from PVC-U and PVC-M fracture tests indicates that different failure mechanisms occur in each case. Figure 3.3 shows failure surfaces after C-ring fracture tests (ISO/DIS 1999) conducted on PVC-U and PVC-M. In PVC-U, the whitened region ahead of the notch corresponds to slow crack growth through a crazed zone, until an abrupt transition in failure mode occurs at a critical crack length (approximately 4.1 mm in this case). The subsequent non-whitened surface corresponds to crack acceleration and fast brittle fracture. In contrast, the PVC-M specimen has failed by slow crack growth in the post-yield regime (indicated by the enhanced whitening on the surface), and eventual plastic collapse of the specimen ligament. The absence of purely brittle fracture in the PVC-M specimen is consistent with the addition of a ductility-promoting modifier.

An increased toughness is exhibited in those materials which can absorb energy during the fracture process (Bucknall 2000). The mechanism that promotes toughness is thought to be initial strain softening and yielding, followed by eventual strain hardening of large volumes of material. This latter strain hardening process is essential, as fracture will occur at a relatively early stage of deformation in its absence. Furthermore, to achieve “tough” extruded pipes, this process must also take place at the tip of any inherent flaws and surface defects in the pipe wall.

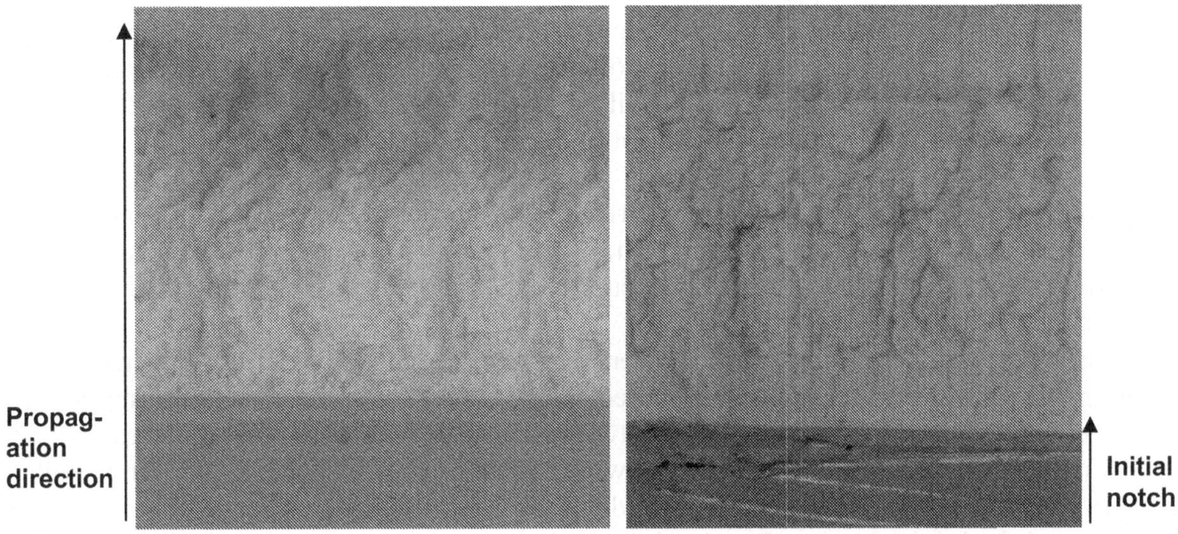

Figure 3.3 Failure surfaces from C-ring tests on PVC-U (left) and PVC-M (right)

Two distinct failure mechanisms have been identified for polymers – multiple crazing, and shear yielding (Whittle 2002). While these mechanisms can occur independently or together, higher levels of toughness are achieved when shear yielding predominates. This ensures that larger volumes of material in the vicinity of the crack tip partake in the strain softening/hardening process.

According to Whittle (2002), the optimization of elastomeric toughening requires the elastomeric particles themselves to firstly undergo strain softening. This is achieved by cavitation of elastomeric particles, which weakens their resistance to deformation, and initiates yielding in the surrounding thermoplastic matrix at reduced stresses. Drawing of the matrix can then occur despite the constraint normally associated with the crack tip. A degree of strain hardening in the elastomeric particles is then induced by stretching of fibrils within cavitated particles. This mechanism explains why the toughening process requires the presence of elastomeric particles (and not simply voids) in the thermoplastic matrix, and an affinity between these particles and the surrounding matrix. In the case of PVC pipes, it is recognized that voids adversely influence long-term performance, and a number of tests, such as high-temperature tests, have been used to detect their presence (AS/NZS 1477). Furthermore, the use of particles which have no chemical affinity for PVC (such as styrene–butadiene–rubber (SBR), for example) does not increase toughness. While SBR particles simply fill voids, CPE has an affinity for PVC and increases toughness by the mechanism described above.

Analysis Methods for PVC-M Pipe

According to Marshall, Pearson and MacKellar (1996), defects can influence failure mechanisms in plastic pipes in two ways:

- A stress concentration is produced at the defect tip, which causes crack growth at relatively low applied stress levels. This will eventually produce brittle fracture failure as described previously.
- The presence of a defect or growing crack reduces the cross-section of the pipe wall. Since less intact material is able to sustain applied pressure, the stress required to cause failure will decrease. This will eventually produce yielding and plastic collapse of the remaining uncracked ligament in the pipe wall.

The failure mode for PVC-M is clearly ductile and proceeds by yielding and plastic collapse, and subsequent ductile tearing. The "competition" between these mechanisms is described in terms of a "net section yield locus" (Marshall, Pearson and MacKellar 1996), as shown schematically in Figure 3.4.

LEFM theory indicates that the dependence between remote failure stress and crack size is non-linear (i.e., the first mechanism described above). In contrast, for uniaxial tension (or circumferential hoop stress), the dependence between failure stress and crack size is linear (i.e., the second mechanism described above). As shown schematically in Figure 3.4, different failure mechanisms will be observed depending on which is lower: the net section yield stress, or the stress required for brittle fracture. For example, while net section yielding may be observed for small cracks, there is clearly a critical size above which brittle failure is observed (point a_c in Figure 3.4) (Marshall, Pearson and MacKellar 1996). For larger crack sizes, ductile failure could again be observed if the net section yield stress falls below the stress required for brittle fracture.

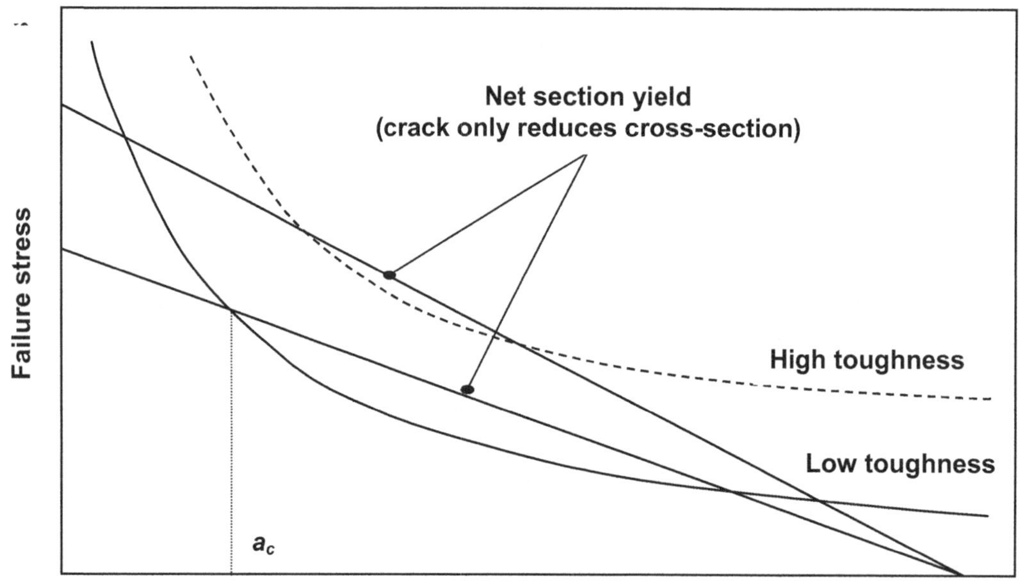

Figure 3.4 Effect of increasing toughness on remote stress required to cause failure

While brittle fracture will be observed for failures controlled by the stress concentration, if the decrease in net section stress is sufficient, the remaining ligament will yield and failure will be ductile. It follows that the failure mechanism observed in practice depends on the toughness and strength of the pipe material, as illustrated in Figure 3.4. If the yield strength of a polymer is increased without a commensurate increase in toughness, then the dominant mode of failure will be by brittle fracture. For PVC-M, failure occurs by a ductile mechanism and can be attributed to the addition of CPE, which increases toughness and reduces yield strength (Whittle and Burford 1998; Marshall, Pearson and MacKellar 1996).

Due to the ductile failure mechanisms of PVC-M, obtaining plane-strain K_{IC} values is difficult, and the applicability of LEFM theory is unclear. However, to solve this problem, fracture in the post-yield (brittle failure) regime can be described using corrections such as that proposed by Bilby, Cottrell and Swinden (BCS) (Whittle, Burford and Hoffman 2000). The BCS yield theory attempts to account for plastic deformation during fracture, and allows plane-strain K_{IC} values to be calculated indirectly (Davis, Burn and Whittle 2001). By including the BCS plasticity correction in Equations 3.1–3.6, a modified LEFM approach can be adopted to predict time to failure for PVC-M with some success.

As stated previously, when either the yield strength has been reduced or the toughness increased sufficiently to promote ductile failure, the validity of LEFM theory is unclear. For ductile tearing, a large plastic zone often exists at the tip of a crack, and much of the plastic flow work that is dissipated in the zone is not directly associated with the fracture process. An alternative fracture property for characterization is the energy per unit area of crack growth, or the *J*-crack growth resistance (Marshall, Pearson and MacKellar. 1996). While the *J*-crack growth resistance indicates the extent of energy absorbed during ductile tearing, its use in predicting time to failure has not been investigated to an extent that would allow it to be used to predict failure.

A second alternative to characterize fracture behavior in the ductile regime is the essential work of fracture (EWF), using double edge notch tensile (DENT), samples as proposed by Broberg (1968). This approach assumes that in quasi-static crack growth, the non-elastic region at the tip of a crack may be divided into two regions: an inner region where the fracture process takes place, and an outer region where plastic deformation takes place (Figure 3.5).

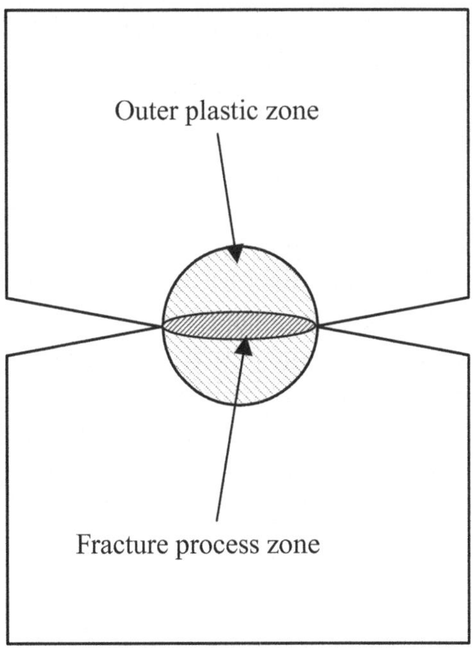

Figure 3.5 Schematic illustration of fracture process zones for a DENT sample

The total work of fracture (W_f) can then be partitioned into two components: work that is expended in the inner fracture process zone to form a neck and subsequent tearing, and work that is consumed in the outer plastic deformation zone (yielding). The former is referred to as the essential work of fracture (W_e) and the latter as the non-essential work of fracture (W_p). W_f can by expressed as:

$$W_f = W_e + W_p \tag{3.8}$$

W_e is proportional to the length of the ligament, and the total W_p is proportional to the square of the ligament (Mai 1986). Equation 3.8 can therefore be expressed as:

$$W_f = w_e lt + \beta w_p l^2 t \tag{3.9}$$

where l is the sample ligament length, t is the sample thickness, β is a geometry-dependant factor related to the plastic zone, w_e is the specific essential work of fracture, and w_p is the specific non-essential work of fracture.

Dividing Equation 3.9 by the cross-sectional area of the ligament lt gives the specific total work of fracture, w_f, as:

$$w_f = w_e + \beta W_p l \tag{3.10}$$

Therefore, if Equation 3.10 can be applied, plotting the variation of w_f with l should give a straight line, with an intercept w_e and a slope equal to βw_p.

By using thin specimens, the EWF method has recently been used to determine the variation in plane-stress K_{IC} through the thickness of PVC-U pipe (Lau and Truss 2002). While this allows ductile tearing criteria to be determined, using EWF for lifetime prediction requires further work.

FRACTURE IN PVC-O PIPES

PVC-O pipes are manufactured with a preferential molecular orientation in the hoop (or circumferential) direction. This orientation provides an increase in strength in this direction. There are two commercial processes for producing PVC-O pipes. The first is an "offline" process in which a small diameter thick-walled conventional PVC-U pipe is extruded. This is then reheated to a temperature between the glass transition and 120°C, and inflated to the desired size. While this is intended to create unaxial orientation in the hoop direction, some axial alignment of molecules is also produced, since the pipe ends are fixed during inflation. The second process is an "inline" process in which conventionally produced pipe is expanded during the extrusion process (Chapman and Ågren 1998). Since the pipe can be drawn simultaneously in the axial direction during inflation, the degree of biaxial orientation (in the axial and hoop directions) can be controlled.

Previous research has indicated that PVC-O pipes do not fail by radial cracking through the pipe wall thickness (as with PVC-U), but rather by crack propagation in the tangential direction around the pipe circumference (Kwon and Truss 2002).

Analysis Methods for PVC-O Pipes

Previous attempts to use three-point bend fracture testing to propagate cracks through the thickness of PVC-O pipes have been unsuccessful, due to the tendency for circumferential cracking (Kwon and Truss 2002). Consequently, this has led to the use of thin specimens machined from the pipe wall, and EWF methods to assess fracture both parallel and perpendicular to the direction of molecular orientation. Comparative testing indicates that EWF is higher for cracks propagating perpendicular to the direction of molecular orientation (Kwon and Truss 2002). This is intuitive, since stronger polar bonding would exist across the path of crack propagation in this case.

As with PVC-M, EWF methods can be used to determine the criteria for fracture, but predicting lifetimes for PVC-O pipes requires further work.

FATIGUE FRACTURE IN PVC PIPES

The design issue of fatigue fracture in PVC pipes has been the subject of extensive research both in Australia and the UK. While it is established that cyclic loading conditions could produce premature failure, it should be noted that there is some doubt whether field failures in PVC pipes are attributed to fatigue (Brogden 2000).

While the term "fatigue" is generally restricted to metals in cyclic loading conditions, plastics also exhibit a reduced capacity to withstand static stresses with increasing time under cyclic loading. Using a square waveform, Gotham (1969) compared the total time under stress for cyclically and statically stressed samples of PVC. An even greater reduction in load-bearing capacity over time was observed under cyclic loading compared to static loading. This confirmed that PVC exhibits a different response to cyclic loading, which must be taken into account when a pipe is subjected to cyclic loads in service. In contrast to fracture surfaces from failure under static load, fatigue fracture surfaces can exhibit intermittent radial "striations" emanating from the point of crack initiation (Elinck, Bauwens and Holmes 1971). While these striations do not coincide with each load cycle, they do indicate "stop-start" crack growth and their total number correlates approximately with the total number of applied load cycles (Elinck, Bauwens and Holmes 1971). According to Joseph (1984), ductile failures are observed in PVC-U when cyclic loads of high amplitudes are used. As the amplitude of applied loading decreases, fracture surfaces become increasingly characteristic of fatigue, with a smooth brittle appearance, together with striations. This suggests that, under high loads, a static failure mode intervenes before sufficient fatigue cycles can be applied (Joseph 1984).

The response of PVC to cyclic loads was investigated by Benham and Hutchinson (1970) using a square waveform with equal tensile and compressive stress components. Again, the time to failure reduced when PVC was subjected to a cyclic load compared to a static load of the same magnitude as the maximum tensile stress. While the time to failure increased for greater cycle time periods, the actual number of cycles required for failure decreased.

The general relationship between the applied SIF range (ΔK) and the crack growth rate per cycle (da/dN) is illustrated schematically in Figure 3.6.

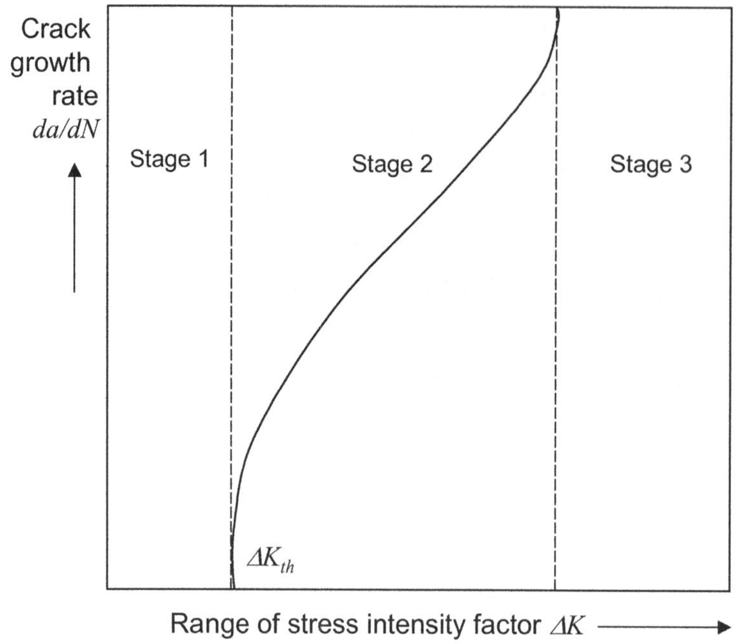

Figure 3.6 Schematic illustration of the relationship between applied SIF range and crack growth rate per cycle

As shown, three regions are identified. In Stage 1, where the SIF amplitude is sufficiently small, there is no crack growth. The transition to Stage 2 is marked by a threshold SIF range, ΔK_{th}, beyond which a crack initiates and proceeds to grow. At some point, the length of this crack will be sufficient for brittle fracture to occur, signified by the rapid upsweep in crack growth rate in Stage 3.

In stage 2, the rate of crack propagation per cycle (*da/dN*) is related to the amplitude of the applied SIF according to the Paris equation:

$$\frac{da}{dN} = A\Delta K^n \tag{3.11}$$

where ΔK is the amplitude of the applied SIF cycle, and A and n are constants. Under cyclic loading conditions, the LEFM definition of an applied SIF in Equation 3.1 is recast as:

$$\Delta K = Y\Delta\sigma\sqrt{\pi a} \tag{3.12}$$

where $\Delta\sigma$ is the amplitude of the applied stress cycle. Substituting Equation 3.12 into Equation 3.11, and integrating with respect to the crack length between the initial length (a_0) and the critical length required for brittle fracture (a_c), gives an expression for the number of cycles to failure N_f:

$$N_f = \int_{a_0}^{a_c} \frac{1}{A\Delta K} da \tag{3.13}$$

While this integral can be solved directly to estimate lifetime, Burn (1994) provided a more accurate algorithm to simulate incremental fatigue crack growth, and account for the change in fracture toughness and LEFM geometry factor over time.

In a comprehensive study of fatigue in PVC, Radon (1977) developed a universal relationship between crack growth rate and applied SIF range:

$$\frac{da}{dN} = F\phi^{\alpha} \tag{3.14}$$

where ϕ is a function of the critical, maximum, and threshold SIFs, and F and α are numerical factors. Various threshold SIFs for PVC-U have been proposed in the literature, ranging from $\Delta K_{th} = 0.24$ MPa $m^{0.5}$ (Burn 1994), $\Delta K_{th} = 0.32$ MPa $m^{0.5}$ (Radon 1977), and $\Delta K_{th} = 0.3$ MPa $m^{0.5}$ (Mills and Walker 1976). It should also be noted that other researchers have reported no threshold SIF at all (Joseph 1984), but it is debatable whether the applied SIF range was sufficiently low (Whittle 2002).

Factors Affecting Fatigue Performance

A review of the literature indicates that there are various factors that influence fatigue performance. These are summarized below.

Stress Amplitude

Results from a recent study in Australia (Whittle 2002) indicate that for specimens of the same ligament thickness, an increase in pressure amplitude significantly reduces the number of cycles to failure under cyclic loading.

Vinson (1981) tested a large number of PVC-U pipes under cyclic loading. A power-law relationship was established between the peak stress and the number of cycles to failure. However, the minimum stress was comparatively low and the stress amplitude was not significantly lower than the maximum stress. Consequently, Vinson (1981) acknowledged that extrapolation of the relationship to very low stress levels was questionable.

A design approach for PVC water pipes, subjected to cyclic loading, was developed by Stapel (1977). The salient features of the design were that the maximum stress should be limited to 12.5 MPa (which is of the order of the static design stress for PVC-U), and that the stress amplitude should not exceed 3 MPa. This stress amplitude was defined as half the difference between the minimum and maximum stress.

Joseph (1984) re-examined published fatigue data for PVC-U pipes and established a lower bound that was subsequently incorporated into the Australian design philosophy (see Appendix A). Accounting for the different definitions of stress amplitude (in Australia it is defined as the difference between maximum and minimum stress), Joseph's proposal is similar to that of Stapel (1977), with a recommended stress amplitude of 5.5 MPa.

Mean Stress

In a study of Australian PVC-U pipes, Kim et al. (1988) demonstrated that fatigue crack growth rates depend on the mean applied stress intensity factor K_{mean} as well as the applied SIF range ΔK. This led to a modified Paris equation for crack growth written as

$$\frac{da}{dN} = B(K_m \Delta K)^n \tag{3.15}$$

where, as before, B and n are constants. In a review of literature up to 1980, Hertzberg and Manson (1980) found that higher mean stresses increase crack growth rates, due to creep extension at the crack tip. However, strain-induced modifications to polymer structure can also occur, which resist fatigue crack growth. As outlined by Burn (1994), this has led to conflicting results for PVC-U where crack growth has been found to both increase and decrease for higher mean stresses.

In a study conducted in the USA, by Jeffrey, Moser and Folkman (2004) compared the observed number of cycles to failure on a set of USA PVC pipes with predictions based on previously published data in the USA and UK. Since predicted lifetimes were extremely conservative, it was proposed that the fatigue lifetime (defined as the number of cycles to failure) of PVC pipe under internal pressure not only depends on the applied stress amplitude, but also the mean stress. Based on apparent similarities between the physical properties of PVC and Steel at high temperatures, mean stress was also proposed to influence fatigue lifetime, especially for high mean stresses or low stress amplitudes (Jeffrey, Moser and Folkman 2004). By plotting experimentally-determined fatigue lifetimes on a graph of mean stress vs. stress amplitude, lifetime curves were generated for different combinations of mean stress and stress amplitude. As a first approximation, it was assumed that similar fatigue lifetimes could simply be joined by a straight line. While empirical equations relating fatigue lifetime to stress amplitude and mean stress were derived, further experimental data is required to validate this approximation (Jeffrey, Moser and Folkman 2004).

Stress Intensity Factor

In an Australian study, Whittle (2002) plotted the variation of the number of cycles to failure against the SIF range ΔK and the product of this range with the mean SIF K_m. While reductions in both these parameters increased the cycles to failure, different sensitivities to pressure amplitude were observed, suggesting that the mean SIF may be influential (Whittle 2002).

Test Temperature

Gotham and Hitch (1978) recognized that temperature was not necessarily constant for PVC pressure pipes in service. Strips of PVC-U were tested under cyclic loads at temperatures of 5°C (41°F) and 23°C (73°F), confirming that the stress required to cause failure in a given time was less under cyclic loads than static loads. The temperature sensitivity of these results did not appear large at the lower stress ranges, and it was concluded that in the UK (with water temperatures between 5°C (41°F) and 8°C (46°F)), temperature would have no adverse effect.

In contrast, Brogden, Shepherd and Marshall (1989) state that changing test temperature can significantly influence the performance of PVC-M pipe. Reducing temperature from 30°C (86F) to 5°C (41F) produced a significant increase in the number of cycles to failure. Whittle (2002) reported similar temperature dependencies for PVC-M and PVC-U pipes. Again, a reduction in test temperature from 30°C (86°F) to 5°C (41°F) significantly increased the number of cycles required for failure.

Processing

Gotham and Hitch (1978) demonstrated that plain samples of well-processed PVC performed better than samples of poorly processed material under cyclic loading. However, in the case of sharply notched specimens, this sensitivity to processing was not evident. It was concluded that different processing (manifested as the quantity of broken down resin grains) affects the crack initiation phase (Stage 1 in Figure 3.6), rather than crack growth (Stage 2 in Figure 3.6). As outlined by Whittle (2002), the larger the inherent flaw size in the residual grain structure of poorly processed PVC, the larger applied SIF compared to well-processed material. By artificially introducing a sharp notch, this difference is eliminated.

Defects and Defect Location

In a study by Vinson (1981), a large proportion of PVC-U pipe samples had scratches that were between 0.13-mm (0.005 inches) and 0.5-mm (0.002 inces) deep and 600-mm (23.6 inches) long, cut in the external surface of pipes at the position of minimum wall thickness. For pipes with scratch depths up to 5% of the wall thickness, the same fatigue performance was achieved as undamaged pipes. Below this depth, failures were also just as likely to occur at the scratch location as anywhere else in the pipe. However, as notch depths increased to 9% of the pipe wall thickness, all failures were observed to occur at the scratch location. Furthermore, the fatigue resistance was reduced, indicating that the influence of defects was dependant on their depth.

Results from tests conducted by Joseph and Leevers (1985) indicated that fatigue cracks in pipes initiated at weld lines associated with the spider in the pipe die. However, as observed by Whittle (2002), it is unclear whether this is a universal effect, or is attributed to the tooling used to manufacture those particular pipes. For example, when testing large diameter pipes, Vinson (1981) observed that 50% of failures occurred at the location of minimum wall thickness (i.e., maximum stress), and the remaining 50% occurred in random locations.

Christie and Phelps (1984) demonstrated that pipes with severe internal notches performed less well than unnotched pipes, when subjected to point loads intended to simulate poor pipe installation. While the results demonstrate that point loads and large internal surface defects should be avoided, they are not significant to well-produced and installed pipes.

An analysis of the extent of external damage incurred in handling PVC pipes was conducted by Burn (1993). Sharp scratches with depths between 0.3 to 0.5 mm were reported, along with occasional deeper defects. The fatigue behavior of PVC-U was then investigated using both compact tension specimen and externally notched pipes. Sharp external notches in pipes reduced the number of cycles to failure to a greater extent than blunt notches when subjected to a pressure regime of 1.2 ± 0.3 MPa. Invariably a crack initiated from the root of the notch and propagated through the pipe wall to failure (Burn 1993). While this suggests that external surface defects may initiate failure in service, Brogden (2000) reported that for PVC-U pipes "the origins of service failures are always found on the pipe bore".

This evidence above suggests that while external surface defects influence the fatigue performance of PVC pipes (with failure time decreasing as defect size increases), field failures under static loading conditions do not usually initiate at the outer surface of the pipe. One reason for this difference may be the fact that circumferential residual stress (created from non-uniform cooling during pipe manufacture) is usually compressive at the pipe outer surface. This would impede crack propagation should it initiate from the pipe outer surface. However, further investigation of field fracture failures is required to confirm whether crack initiation and propagation can occur from the pipe outer surface in the field.

Fatigue Performance of PVC-U, PVC-M and PVC-O Pipes

The fatigue behavior of PVC-U and PVC-M pipe formulations was compared by Fitzpatrick et al. (1996). The PVC-M resin contained CPE concentrations of 5–25 pphr. A progressive decrease in fatigue resistance was observed with increasing CPE concentration, and there was no apparent fatigue threshold. However, it is questionable whether the minimum applied stress amplitude was sufficiently low to observe a threshold if one existed. Also, the discrepancy between the PVC-M sample phase morphology and the phase morphology of the PVC-M pipe used by Fitzpatrick et al. (1996) has been highlighted (Whittle 2002).

Lawrence, Teo and Potter (1998) compared the temperature sensitivity of PVC-U and PVC-M under cyclic loading. While a small temperature effect was observed for PVC-U (consistent with previous observations (Hertzberg and Manson 1980)), the temperature sensitivity of PVC-M was much greater. An equivalent performance for the two materials was observed at approximately 20°C (68°F).

By applying a rotational bending stress to circumferentially notched rods at 4.6 Hz, Brogden (2000) compared the fatigue behavior of PVC-U and PVC-M. The results were plotted as the variation in cycles to failure with stress range. Above a range of approximately 5 MPa (725 psi), the lines of best fit for the two materials almost coincided. Below this stress range, PVC-M samples exceeded the number of cycles to failure anticipated by extrapolation.

In a recent Australian study (Whittle 2002), the fatigue resistance of PVC-U and PVC-M samples was similar, with both exhibiting a fatigue threshold, when tested by rotational bending of ΔK_{th} = 0.3 MPa m$^{0.5}$. In contrast to Whittle (2002), Brogden (2000) applied a factor of 1.5 to the tensile stress amplitude. The temperature sensitivity of the fatigue resistance of these materials was also reported to be similar (Whittle 2002). These measurements were performed on Ca/Zn stabilized PVC-U compared to Burn's (1994) work, which was carried out on Pb-stabilized pipe and may indicate that Ca/Zn pipes have a superior fatigue performance to Pb-stabilized pipes. This effect is confirmed by unpublished fatigue studies carried out by the authors.

In comparison to PVC-U and PVC-M, only limited fatigue performance data have been published for PVC-O. As stated by Dukes (1985), the dynamic fatigue performance of PVC-O pipe "is always better than extruded pipe, when results are obtained by the same means". When tests are performed using ring specimens, PVC-O samples failed after 300,000 cycles compared to 140,000 for PVC-U (Dukes 1985).

CHAPTER 4
LITERATURE REVIEW AND REPORT – PERFORMANCE STANDARDS FOR PVC PIPE

INTRODUCTION

The term "PVC" is a common nomenclature that can specifically include unplasticised PVC (PVC-U), modified PVC (PVC-M) and oriented PVC (PVC-O). Additionally PVC-M is a generic term, which is used to describe a PVC material with a wide range of modifiers in its formulation, such as chlorinated polyethylene, to improve ductility. To complicate standardization further, all PVC pipe compounds must be stabilized against heat degradation, and may need to be stabilized against UV degradation to protect against prolonged outdoor or unshaded storage in southern parts of the USA. For heat stabilization, a range of systems such as lead, tin, calcium/zinc and the new organic stabilizers are available, While for UV stabilization, TiO_2 is commonly used in countries such as Australia, the USA and Canada.

While current performance standards indicate that water utilities should expect a minimum 50-year service life from a PVC pipe, expert opinion suggests that 100 years is a conservative estimate for a "properly designed and installed pipe" (Uni-Bell 2002). In order to achieve these lifetimes, valid performance requirements are needed in manufacturing standards to ensure that pipes are manufactured to a suitably high standard. However, despite the availability of performance standards in countries such as the United Kingdom and Australia, premature failure of PVC pipelines has occurred, and consequently standards in these countries are being continuously upgraded to minimize the probability of future failures. PVC pipelines can fail in a number of ways, associated with heat degradation, UV degradation, slow crack growth, and fatigue. The standardization process needs to consider these failure mechanisms and ensure that sufficient performance tests are incorporated into standards to maximize lifetimes.

Standardization of PVC pipes in the USA and Canada is covered by a plethora of standards, and includes AASHTO, ANSI, ASAE, ASTM, AWWA, BOCA, and many other federal bodies (Uni-Bell 2001). However, PVC pressure pipes are generally manufactured and tested in accordance with either ASTM or AWWA standards. The "rural" water market (i.e., rural areas and smaller towns/villages) will usually specify PVC pipe products manufactured to ASTM standards (ASTM 2241). Pipes manufactured to this ASTM pipe standard have outer diameters that are compatible with those of steel pipes. The "municipal" pipe market (i.e., generally larger towns and municipalities) usually requires that PVC pipes be manufactured to one of the several AWWA pipe standards. For these standards, the outer diameters are compatible with ductile iron pipes and fittings. These pipes are often required to provide fire protection in addition to potable water supply, and therefore may also be certified by a third-party testing laboratory, such as UL, FM, CSA, or ULC. The requirements of these testing/certification organizations are typically very similar to those defined in the AWWA PVC pipe product standards. Consequently, this comparison of existing manufacturing, installation and performance standards for PVC pressure pipes in North America, Europe, and Australia is confined to AWWA and ASTM standards.

MATERIAL SPECIFICATION

Material specifications for the manufacture of PVC pipe in the USA are more rigorous than other standards (such as ISO), as indicted in Table 4.1.

Table 4.1
PVC resin specifications for different standards

Standard	Pipe type	PVC material classification	Relevant standard
ANSI/AWWA C900-97	PVC-U	12454	ASTM D1784
ANSI/AWWA C905-97	PVC-U	12454	ASTM D1784
ANSI/AWWA C909-02	PVC-O	12454	ASTM D1784
ASTM D1785-99	PVC-U	12454	ASTM D1784
	PVC-M	14333	
ASTM D2241	PVC-U	12454	ASTM D1784
	PVC-M	14333	
ASTM F1483-98	PVC-O	12454	ASTM D1784
		12454	
AS/NZS 1477	PVC-U	Not defined: 1.5 pphr TiO_2, min. of 83.3% PVC, Vicat ≥80°C	
AS/NZS 4765 (Int)	PVC-M	Not defined: 1.5 pphr TiO_2, min. of 83% PVC, Vicat ≥80°C	
AS/NZS 4441 (Int)	PVC-O	Not defined: 1.5 pphr TiO_2	
ISO 4422-1996	PVC-U	Not defined	
ISO/DIS 16422	PVC-O	Resin >K64, Vicat ≥80°C	

By calling up ASTM D1784, the pipe manufacturing standards effectively specify a minimum impact, tensile modulus of elasticity, and deflection temperature. The class requirement most commonly specified of 12454 would be easily met by most PVC resins used for manufacturing PVC-U pipes, and even compounds with significant levels of regrind would meet these requirements. However, specifying this grade of resin would not guarantee the manufacture of pipes with satisfactory lifetimes, as resins meeting these requirements are known to have performed poorly in countries such as the United Kingdom and Australia. One exception is the option of using resin grade 14333 in manufacturing pipes to ASTM D1785 and D2241. Specification of this grade means that the impact resistance of the compound has to be increased at the expense of the modulus of elasticity. This is commonly done by adding impact modifiers such as acrylics and CPE, and generally these pipe types are classified as PVC-M pipes to distinguish them from PVC-U pipes. These additives have the effect of increasing polymer ductility, and ensure that PVC pipes are more resistant to failure under slow crack propagation from point loading sources. However, at the time of writing this report, it is understood that no USA or Canadian manufacturers are known to be using the 14333 resins in the manufacture of PVC pipe (Uni-Bell, pers. comm.). The practice of specifying the compound class has some merit, however, it does inhibit manufacturer innovation, is a prescriptive practice, and, probably more importantly, can be modified significantly by manufacturers that blend their own feedstock. Consequently, the specification of the PVC resin grade may have only a small effect on the final or field performance, which is controlled by the following factors:

- Hydrostatic pressure resistance.
- Slow crack growth resistance.
- Installation impact requirements.

- Resistance to heat and UV radiation.
- Manufacturing competency.
- Effect on water quality.

While the PVC compound has an effect on each of these properties (manufacturing competency is affected by resin grade), it acts in conjunction with other properties such as pipe dimensions (impact) and processing history (heat and UV). A better practice that would guarantee adequate field performance would be to ensure that relevant performance tests are included within the manufacturing standards and installation codes. This would remove prescriptive requirements and allow the manufacturers to innovate fully on resin development. Some of these tests are discussed below.

HYDROSTATIC PRESSURE RESISTANCE

The response of PVC pipe to hoop stress induced by internal pressure is time dependant, with the stress levels required to cause ductile failure decreasing with time. The stress/time failure curves developed using long-term hydrostatic design stress testing are commonly used in designing PVC pipes, in the belief that this will safeguard the pipes against premature failure. However (as discussed in Chapter 3), field failures rarely occur in the ductile failure manner predicted by the hydrostatic regression curves (a possible exception may be PVC-M, however this is still to be shown), and occur by slow crack growth followed by fast brittle fracture, which depends on the operating stresses, the processing level of the compound, and the fracture toughness of the pipe material. However, the long-term hydrostatic design process ensures that the pipe is operating at a low enough stress level so that ductile failures will not occur, and the possibility of brittle failures is minimized. To guard against brittle failures, a range of different performance requirements is needed, and these are discussed later. Despite its shortcomings, the method of designing using long-term hydrostatic testing is well entrenched worldwide and would be difficult to change in the short term.

In the USA, long-term hydrostatic design stress (LTHS) testing is conducted in accordance with the Plastics Pipe Institute Technical Report PPI-TR3, where at least 18 pipe failure points are collected, and a regression line is extrapolated out to 100,000 hours. The value at this point is deemed the hydrostatic design basis (HDB) (or LTHS) and, by utilizing a safety factor, F, the hydrostatic design stress of the pipe, S, is calculated according to Equation 4.1:

$$S = \frac{HDB}{F} \tag{4.1}$$

As shown in Table 4.2, the HDB in USA standards for PVC-U and PVC-M resin is 4,000 psi, with standards from other countries being slightly lower. For PVC-O, the HDB for USA standards is 7,100, which includes LTHS values ranging from 6,810 to 7,920 psi. This can be compared to values in other countries varying from 7,245 psi to 4,565 psi.

Table 4.2
Hydrostatic stress design basis

Standard	HDB (psi)	Calculation method for HDB	*S* (psi)	*F**	Surge allowance (psi)
ANSI/AWWA C900-97	4,000	Mean regression†	1,600	2.5	30–40*
ANSI/AWWA C905-97	4,000	Mean regression†	2,000	2.0	None
ANSI/AWWA C909-02	7,100	Mean regression†	2,840	2.5	23–31*
ASTM D1785-99	—	Mean regression†	2,000 1,600 1,250 1,000	— — — —	None
ASTM D2241		Mean regression†	2,000 1,600 1,250 1,000		None
ASTM F1483-98	7,100	Mean regression†	3,550 3,150	2.0 2.25	None
AS/NZS 1477	3,770‡ or 3,420‡	97.5% lower confidence limit regression	1,780‡ or 1,595‡	2.11 2.14	None
AS/NZS 4765 (Int)	3,550	97.5% lower confidence limit regression	2,536	1.4	None
AS/NZS 4441 (Int)	5,942	97.5% lower confidence limit regression	3,420	1.74	None
ISO 4422-1996	3,623	97.5% lower confidence limit regression	1,811‡ 1,500‡	2.0 2.5	None
ISO/DIS 16422	7,245–4,565	97.5% lower confidence limit regression	5,175–2,282	1.4–2.0	

* Associated with a 2 ft/s velocity change.
† Regression of 18 points extrapolated to 100,000 hours according to ASTM D2837.
‡ Based on pipe size; higher value is for thicker pipes.

The HDB depends on a significant number of factors, especially for PVC-O where the extent of molecular orientation and original feedstock can have a significant effect on strength enhancement. Similarly with PVC-M, the method of modification can alter the slope of the hydrostatic regression line and the final HDB obtained. While the values of HDB used and whether different values are used for different resins is not questioned, the methodology of using a standard mean regression is questionable. Hydrostatic testing is an experimental technique where the statistical spread of the data can be significant, depending upon the quality of manufacture. The use of a standard regression does not take this spread of data into account, and it is recommended that a regression technique be adopted that incorporates this potential data spread. ASTM D2837 has gone part of the way to taking data spread into account by requiring the calculation of the lower confidence limit (LCL) and comparison of this value with extrapolated LTHS. However, a recommended option would be to accept international practice, such as detailed in ISO 9080, and to use the 97.5% LCL value as the value for HDB or LTHS.

Water distribution systems are generally designed on an operating pressure or design stress basis. *S* is calculated from the HDB by applying a safety factor or, in the case of AWWA C900 and C909, by applying a safety factor and a surge allowance. The safety factors chosen are variable and dependant on the subjective judgments of the standards writing authority. Table 4.2. summarizes many of the factors used around the world, with values ranging between 1.4 and 2.5 for effectively the same pipe materials. While choosing the safety factor is essentially a national decision, it does have a significant effect on the final design stress under which the pipes operate and therefore, their likelihood to fail under a brittle failure mechanism. As shown in Table 4.3., the final design stresses for pipes manufactured in the USA is not consistent, both within AWWA standards, and between AWWA and ASTM standards, and a case exists for the development of one consistent standard for each of PVC-U, PVC-M and PVC-O.

Table 4.3
Designed operating stresses

Standard	Maximum operating stress (psi)		
ANSI/AWWA C900-97	1,200 (DR25)	1,275 (DR18)	1,300 (DR14)
ANSI/AWWA C905-97	2,000		
ANSI/AWWA C909-02	2,000		
ASTM D1785-99	2,000	Wall thickness controlled series	
ASTM D2241	2,000	SDR controlled series	
ASTM F1483-98	3,550	Resin 1135	
	3,150	Resin 1131	
AS/NZS 1477	1,780	Pipes ≥7 inches	
	1,595	Pipes ≤6 inches	
AS/NZS 4765 (Int)	2,536		
AS/NZS 4441 (Int)	3,420		
ISO 4422-1996	1,811	Pipes ≥4 inches	
	1,500	Pipes ≤3.5 inches	
ISO/DIS 16422	5,172–2,282	Depends on manufacturing process and safety factor	

The choice of safety factors can be subjected to a more scientific approach, and Table 4.4 shows one such approach that has been applied for water distribution main rehabilitation with polyethylene pipe. The application of a scientific approach accounting for manufacturing and installation conditions would be a major advance on the current, seemingly arbitrary methods of setting these factors.

SURGE AND FATIGUE

While the safety factors chosen within the USA are comparable with those used worldwide (although they are significantly higher than those recommended by ISO), probably the most contentious factor is the use of a surge allowance in AWWA C900 and C909. The inclusion of a surge allowance has the effect of significantly lowering the design stresses, to a point where they are significantly below those used elsewhere in the world, as seen in Table 4.3. This practice adds significantly to the cost of PVC pipe, since pipe wall thickness is required to be up to 30% thicker to operate under a similar operating pressure (Equation 4.2):

$$P = \left(\frac{2}{\frac{D}{t} - 1} * \frac{HDB}{F} \right) - P_S \qquad (4.2)$$

Table 4.4
Design factors for water and sewerage pipelines

Condition	Installation	Factor	Index
Fluid	Water	1.0	f_0
	Domestic sewage	1.0	
	Industrial sewage	1.2	
Soil, fluid, or pipe temperature (average *t*°C)	$-20 < t \leq -10$	Refer manufacturer	f_1
	$-10 < t \leq 0$	0.6	
	$0 < t \leq 20$	1.0	
	$20 < t \leq 30$	1.1	
	$30 < t \leq 35$	1.25	
	$35 < t$	Refer manufacturer	
Location based on minimum depth of cover	Open field	1.0	f_2
	Minor country road shoulder	1.0	
	Major country road shoulder	1.0	
	Minor country road – under pavement	1.1	
	Major country road – under pavement	1.2	
	Residential – paved and unpaved nature strip (footpath)	1.0	
	Residential roadway — under pavement	1.1	
	Major urban road – under pavement	1.2	
	Commercial/industrial – paved and unpaved nature strip (footpath)	1.1	
	Commercial/industrial roadway – under pavement	1.2	
	Central business district	1.4	
	Private land – easement	1.0	
	Above ground	1.0	
	Submarine crossings	1.4	
Installation method	Standard trenching	1.0	f_3
	Plough-in	1.1	
	Directional drilling	1.2	
	Slip line with back grouting	1.0	
	Slip line without back grouting	1.2	
	Pipe cracking – with liner pipe in situ	1.0	
	Pipe cracking – with liner pipe removed	1.2	

Notes:
Choose only one factor from each condition. Pumped installations require further design consideration.
Design factor $F = f_0 \times f_1 \times f_2 \times f_3$.
Example: For a water reticulation pipeline in a commercial paved mall used primarily for pedestrian traffic, operating at maximum summer water temperature of 24°C, and laid using pipe cracking with liner pipe removed: $f_0 = 1.0$; $f_1 = 1.1$; $f_2 = 1.1$; $f_3 = 1.2$.
Design factor $F = 1.0 \times 1.1 \times 1.1 \times 1.2 = 1.5$.

The practice of subtracting a surge allowance from the design pressure of the pipes is questionable, and for most pipelines operating under static pressure, it is not needed. Scientific evidence would suggest that for the occasional short-term overpressurization, the pipe would cope satisfactorily as discussed by Marshall and Brogden (1997) whose work lead to the publication of new design practices for the United Kingdom (Water UK Ltd 2000). The mechanical properties of PVC, like all other polymers, are time dependant and, under surge conditions, stiffness increases and a pipe can sustain the associated additional load (Marshall and Brogden 1997). However, special consideration may be needed when the pipe is constrained by tapping bands, flanges, fittings, etc., where brittle failure may be a concern. The demonstrated ability of plastics pipes to accommodate short-term surge pressures can be utilized to advantage by allowing some overpressurization for the pipe. However, for fittings, generally the safest approach would be to design so that the peak pressure for the fittings does not exceed the field test pressure of the pipe, which is normally 1.25 times the maximum system operating pressure. For example, a system operating normally at 100 psi, with occasional surging to 125 psi, will require 100 psi pipe and 125 psi fittings, and system testing should be conducted at 125 psi. A further detailed discussion on this aspect of pipe system design is included in Appendix A.

If occasional overpressurization is irrelevant and the design issue is actually cyclic pressures, then the pipeline should be designed for fatigue conditions, and the application of a surge allowance is a simplistic approach. Fatigue is a slow crack growth phenomenon, which is controlled by a number of factors, including the presence of a stress concentrator, the surge amplitude frequency, and mean pressure, as discussed by Burn (1994). Pipes subjected to cyclic loading conditions should be specifically designed for these conditions, and one such process is detailed in Appendix A. In brief, this process requires the number of pressure cycles in service to be estimated, and the operating pressure of the pipe to be de-rated accordingly, as shown in Figure 4.1.

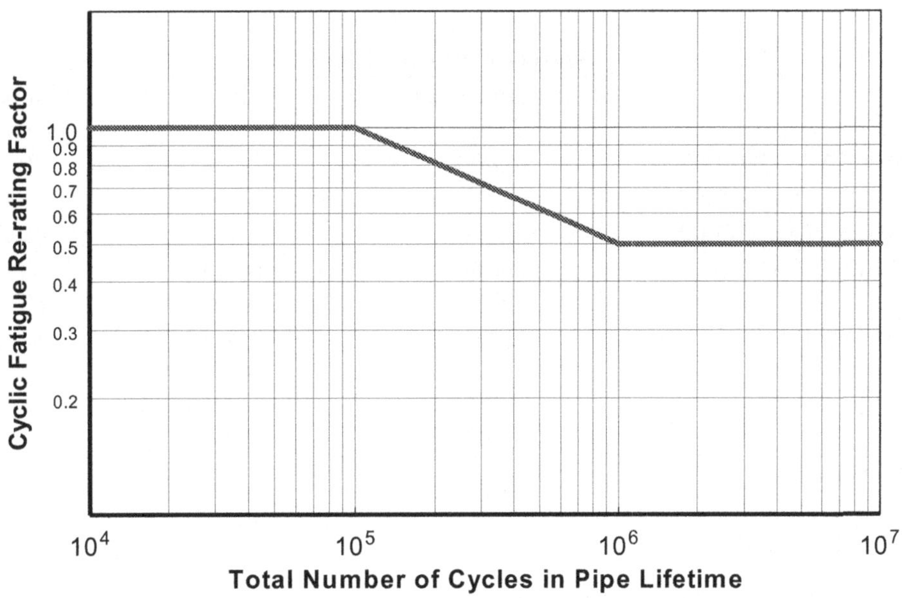

Figure 4.1 Suggested fatigue de-rating factors (see Appendix A)

In addition to cyclic fatigue re-rating factors such as those shown in Figure 4.1., Jeffrey, Moser and Folkman (2004) also provide a design guide for USA PVC pipes subjected to cyclic operating pressures. This design guide makes recommendations based on the applied mean stress and stress amplitude. Although details are not included in this study, a report outlining the theory behind this method and the accompanying design guide (Jeffrey, Moser and Folkman 2004) can be accessed through Uni-Bell (www.uni-bell.org).

As well as design, pipe manufacturing plays a critical role in the ability to meet long-term pressure requirements. Continuous tests are normally specified for the manufacturer to ensure that the ability to meet the HDB is maintained. Generally this is done via either a quick burst test (Table 4.5), a hydrostatic integrity test (Table 4.6), or a short-term sustained pressure test (Table 4.7). All of these tests are prescriptive and do not address the changing requirements of resin selection or the pipe manufacturing process. What these tests need to ensure is that the manufacturing process is of a high enough quality so that the manufactured pipe will continue to meet the long-term hydrostatic performance requirements. This can be achieved by selecting samples on an ongoing basis and testing them to see if they are within the same statistical variation as those that were used to develop the hydrostatic regression line. Unfortunately, this requires the manufacturer to have undertaken sufficient tests to develop this regression line in the first place. This methodology is being adopted throughout the range of Australian standards for pipeline materials, and has been proposed to a range of ISO committees to cope with the varying compounds and manufacturing techniques now associated with PVC pipe manufacture. This is generally proposed as a one-hour test with a hoop stress (*S*) value equal to the 99.5% LCL value at one hour, determined from the 50-year linear regression analysis for the appropriate resin, as detailed ISO 9080. By testing according to this requirement, the variations associated with resin selection and the manufacturing processes are eliminated, and the test ensures that the manufacturer is assessing the ongoing samples to a satisfactory performance standard.

Table 4.5
Short-term burst test

Standard	Short-term test stress (psi)	Resin	Test period (seconds)
ANSI/AWWA C900-97	>6,400	—	60–70
ANSI/AWWA C905-97	—	—	—
ANSI/AWWA C909-02	12,000	—	60–70
ASTM D1785-99*	6,400 5,000	12454-B, 12454-C, 14333-D	60–70 60–70
ASTM D2241*	6,400 5,000	12454-B, 12454-C, 14333-D	60–70 60–70
ASTM F1483-98	11,100 9,900	12454-B (1135)† 12454-B (1131)†	>60 >60
AS/NZS 1477	—	—	—
AS/NZS 4765 (Int)	—	—	—
AS/NZS 4441 (Int)	—	—	—
ISO 4422-1996	—	—	—
ISO/DIS 16422	—	—	—

* Test is optional; the manufacturer may use an accelerated regression test in place of this test and the burst pressure test.

† Designates material designation code.

Table 4.6
Hydrostatic integrity

Standard	Short-term test stress	Test period (seconds)
ANSI/AWWA C900-97	Four times pressure rating	5
ANSI/AWWA C905-97	Two times pressure rating	5
ANSI/AWWA C909-02	Four times pressure rating	5
ASTM D1785-99	–	–
ASTM D2241	–	–
ASTM F1483-98	–	–
AS/NZS 1477	–	–
AS/NZS 4765 (Int)	–	–
AS/NZS 4441 (Int)	–	–
ISO 4422-1996	–	–
ISO/DIS 16422	–	–

Table 4.7
Short-term tests – sustained pressure

Standard	Short-term test stress (psi)	Resin or specific details	Test period (hours)
ANSI/AWWA C900-97	4,200	—	1,000
ANSI/AWWA C905-97	—	—	–
ANSI/AWWA C909-02	6,500	—	1,000
ASTM D1785-99*	4,200	12454-B, 12454-C, 14333-D	1,000
	3,360	14333-D	–
	2,800	14333-D	–
	2,300	14333-D	–
ASTM D2241*	4,200	12454-B, 12454-C, 14333-D	1,000
	3,360	14333-D	–
	2,800	14333-D	–
	2,300	14333-D	–
ASTM F1483-98	7,400	—	1,000
	6,600	—	–
AS/NZS 1477	5,740	Only applicable to pipes with wt <0.16 inches	1
AS/NZS 4765 (Int)	5,507	—	1
AS/NZS 4441 (Int)	6,086	—	1
ISO 4422-1996	5,072	—	100
ISO/DIS 16422	1.65 working pressure		1,000

* Test is optional; manufacturer may use accelerated regression test in place of this test and the burst pressure test.

RESISTANCE TO SLOW CRACK GROWTH AND BRITTLE FRACTURE

A critical factor for PVC that is not addressed in USA standards is a performance requirement to ensure that the pipe has the ability to resist slow crack growth from point sources. This resistance to slow crack growth is a relationship between the fracture toughness of the pipe compound, the yield strength, and the change of these two properties with time. An ideal relationship is high fracture toughness and a low yield strength, which will ensure that ductile failure is the predominant failure mode. Some PE compounds have these characteristics and so tend to fail predominantly in a ductile mode. For PVC, the yield strength is higher than PE, so a trade-off mechanism between ductile and brittle failure occurs, as discussed in Chapter 3. As a consequence, to ensure ductile failure is the controlling mechanism, pipes need to be manufactured with high fracture toughness. This has only been recognized in the last decade or so, and the research results are only now being fed into relevant standards.

Both Australian and ISO standards have a fracture toughness requirement for PVC-U pipes. For PVC-M pipes, a modification of the fracture toughness requirement has been included into the Australian Standard (AS/NZS 4765) to ensure that the failure mechanism always remains ductile. For PVC-O, because of molecular orientation, fracture toughness is difficult to measure using standard tests and is not considered an appropriate test method, however EWF assessment in this area is showing promise, and it is recommended that the standardization committees in the USA become familiar with this work so that its incorporation into USA standards can be considered. Work on EWF is discussed in Chapter 3 and 6. It is recommended that USA standards incorporate a fracture toughness requirement into their standards, in line with ISO recommendations for PVC-U and consider a ductility-type assessment for PVC-M pipes.

Performance analysis (Chapter 5) indicates that the USA has not had the same problems with premature failures compared to other countries. Whether this is due to the pipelines operating under lower stresses, or due to the use of specific formulations or manufacturing procedures which improve toughness, is not known conclusively. As discussed later, Sn-stabilized PVC-U pipe produced by one manufacturer in the USA has been shown to have improved fracture toughness performance compared to Sn-stabilized PVC-U pipe manufactured in Australia. Analysis of the formulations of two typical pipes as detailed in Table 4.8 could provide no valid reason for this and, in fact, based on formulation, the Australian pipe should be tougher because based on the balance of lubricants, especially the external ones, i.e. paraffin waxes, it should fuse faster, at a lower temperature and higher torque and thus have a higher processing level. One concern with this formulation however is that it could be easily over processed.

It is possible that manufacturing issues such as screw design or wear could be the answer to the variation. However, one factor that may influence the better performance of PVC-U pipes in the USA may lie in the selection of the impact test values. By selecting values significantly higher than those set in other international standards, the manufacturing process could well be optimized to produce a better-gelled pipe utilizing a higher processing temperature to meet this requirement.

It is commonly known that the processing temperature can have a significant effect on the pipe properties that are produced (Burgess 1982) with the optimal processing temperature being formulation dependent and it needs to be determined by the manufacturer. For example for PVC-U formulations it is known that there is processing window in which pipe with good gelation is produced in the region of 190°C- 210°C and that the optimal toughness conditions may be achieved with less than 100% gelation (Burgess 1982). If we look at the processing temperatures of the PVC-U specimens, determined from DSC analysis, as detailed in Table 4.9 we can see that the processing temperature of the Australian specimen is much lower than that of the USA specimen which has also resulted in a much lower level of gelation. It is thus possible that the higher impact requirements of the United States Standards have inadvertently led to the production of a tougher pipe, however; this preposition could only be confirmed by an industry benchmarking process to confirm all manufacturers meet the performance requirements.

Table 4.8
Typical pipe formulations for Australia and the USA

	Typical USA Formulation (PHR)	Typical Australian Formulation (PHR)
PVC Pipe Resin (K67)	100.0	100.0
6% Tin pipe stabiliser	0.40	
Calcium Stearate	0.60	0.65
Oxidised Polyethylene Wax 102°C – 0.93 ,Acid No 14-17	0.20	
165 Paraffin Wax, 71°C- 0.91SG	1.75	
1μ Calcium Carbonate	4.00	
Titanium Dioxide	1.50	2.00
Thermolite 178 - Atofina		0.70
Filler 1T - Omya		2.70
Metablen P700		0.75
Loxiol G60		0.15
Paraflint H1-Schuman Sasol		0.50
H12 - Clariant Licolub		0.15
AC617 Wax		0.15
AC307A		0.10

Table 4.9
Process temperatures and gelation levels for USA and Australian PVC-U

Sample Reference	Processing Temperature °C	% Gelation
USA Formulation	189.6	67.8
Australian Formulation	177.3	29.5

Note 1: % Gelation = Endotherm A ÷ (Endotherm A + Endotherm B) where the onset temperature at endotherm A was located between 115 - 118°C and the onset temperature at endotherm B was set at 240°C or peak reaching prior 240°C.
Note 2: The processing temperature is defined as the peak in the DSC curve between the two endotherms, which has been related to manufacturing processing temperature.

The significant variation between these two Sn-stabilized pipes (that should have similar properties) reinforces the need for valid performance requirements such as fracture toughness and gelation. One concern with fracture toughness specification is that specifying an initial value of fracture toughness by itself, does not guarantee long-term performance, because it has been found that pipes with satisfactory fracture toughness levels can have poor levels of gelation. Poorly gelled pipes have been observed to have a steeper fracture toughness/time relationship, which means that the ductile/brittle transition is reached early in their lifetimes, with premature failures initiating from the inside of the pipes. Consequently, as well as specifying fracture toughness, it is also necessary to specify a test to ensure adequate processing has occurred. Many standards have incorporated tests to try and ensure that the level of gelation is adequate. The simplest and oldest, such as those specified in AWWA and ASTM standards (see Table 4.10), are those that test the resistance of the pipe to a solvent such as acetone. The solvent attacks poorly gelled pipes so that surface roughening occurs. Acetone is now considered to be too benign as it only attacks PVC-U of very low gelation levels and has been withdrawn from most international standards. Dichloromethane is now more commonly used and is specified in an ISO standard (ISO 9852). However, this method has disadvantages too, in that judging the degree of attack and hence whether satisfactory processing has occurred is subjective. Nevertheless, its use in preference to acetone assessment in USA standards would be an improvement on the current practice, and adoption of the clauses in the current European standard (EN 1452) may meet the requirement for PVC-U. Unfortunately, dichloromethane can also attack other additives in PVC, and the effect of this must be considered if an improved solvent attack test is considered for adoption in the USA.

Significant work has gone into developing quantitative test procedures, such as capillary rheometry and DSC, to characterize the level of gelation, with DSC showing promise for use as an industry technique, as discussed by Potente and Schultheis (1987) and Teh et al. (1989). However, as yet no standard test method has been developed based upon this method, although there is a subcommittee of ISO TC 138 that is considering the benefits of different procedures. Australia has developed an industry standard to allow optical assessment of the gelation level, and this method (as detailed in Appendix B) is currently being assessed by the Australian industry to determine its suitability for inclusion in national standards. The USA industry may consider such a test method preferable to the use of dichloromethane, where occupational health and safety issues may be of concern.

Table 4.10
Extrusion quality and processing level

Standard	Test method	Required value
ANSI/AWWA C900-97	Acetone	ASTM D2152 <50% attack
ANSI/AWWA C905-97	Acetone	ASTM D2152 <50% attack
ANSI/AWWA C909-02	Acetone	ASTM D2152 <50% attack
ASTM D1785-99	Acetone	ASTM D2152 <50% attack
ASTM D2241	Acetone	ASTM D2152 <50% attack
ASTM F1483-98	Acetone	ASTM D2152 <50% attack
AS/NZS 1477	Fracture toughness	
AS/NZS 4765 (Int)	Long-term toughness	
AS/NZS 4441 (Int)	No test	
ISO 4422-1996	Fracture toughness	<3.25 MP√m depending on wall thickness
	Dichloromethane	No attack
ISO/DIS 16422	No test	

HEAT AND UV RESISTANCE

The stability of PVC formulation under UV degradation and high temperature conditions depends on the stabilization system used, and the level of stabilizers. When PVC is exposed to heat or UV radiation in the 290–400 nm band, a series of complex reactions occurs which cause the PVC to degrade, especially in the presence of oxygen. The accepted theory is that irradiation excites the PVC polymer chains, causing radicals to form, which may further chain-react with the PVC, causing dehydrochlorination, bond cleavage, or crosslinking. The dehydrochlorination results in the formation of highly colored polyenes, which absorb light strongly and may auto-accelerate the dehydrochlorination. The radicals also react with oxygen to initiate chain cleavage or crosslinking, and form the hydroperoxide and carbonyl groups found in degraded polymer. Consequently, stabilizers are needed to protect PVC from heat degradation (thermolysis) as well as light degradation (photolysis). The level of stabilizers added is generally controlled by the thermal performance needs of extrusion, and generally little consideration is given to the long-term thermal performance requirements due to weathering, although for long-term performance at elevated temperatures this needs to be considered.

The existing USA standards for PVC pressure pipes and fittings for pressure applications do not specify requirements for assessing the extent and effect of thermal and UV degradation for PVC pipes. However, both USA and Canadian manufactures routinely add 1.0–1.5 pphr TiO_2 to their pipe and fitting compounds. The only performance requirement in this regard that has any relationship to heat performance is a requirement for a minimum softening temperature or Vicat requirement for the PVC resin. This softening temperature, which is 70°C (158°F) in the USA and 80°C (176°F) in Australia, goes some way to minimizing distortion of pipe when it is stored outside, as pipes will rarely achieve this temperature unless stored in very adverse conditions. However, with high-intensity radiation, under protected conditions and with darker colors, the surface temperature of the polymer can approach the softening temperature, with resultant distortion. Specifying a Vicat temperature does not ensure durability, and minimum requirements for heat and UV stabilizers should be specified. USA (AWWA M23) (AWWA 2002) and Australian/New Zealand (AS/NZS 2032) (Standards Australia/Standards New Zealand 1998) standards do go part way to addressing the heat and UV durability requirements for PVC pipes by specifying storage requirements and defining de-rating of pipes operating in elevated temperature conditions.

When it comes to specifically defining performance requirement for either heat or UV degradation, only Australian standards appear to have considered this issue, and they cover the performance with a prescriptive requirement for 1.5 pphr TiO_2 content. TiO_2 protects the polymer by absorbing a large amount of the UV radiation falling on the polymer, thus greatly inhibiting the formation of radicals and subsequent photo-oxidation, which leads to polymer breakdown. TiO_2 is approximately 50% more costly than PVC resin and its use can lead to increased barrel and screw wear, as well as causing other production problems, thus requiring tighter process control. Although Australian standards do try and address durability issues with the inclusion of TiO_2, there is no assessment of the effects of post-processing thermal degradation and weathering degradation on the long-term mechanical and physical properties, and there is no requirement for post-processing stabilizer additive levels to deal with thermal and weathering degradation.

An extensive literature/standard search failed to identify any alternate standards that assess the effects of thermal degradation and weathering degradation in PVC pressure pipe formulations for water use. However, where PVC is used in cases where outdoor exposure is common (especially PVC windows), and for some non-water pipe applications (i.e., gas) this issue has been partially addressed. Commonly, the requirement for UV durability is that PVC specimens retain 50% of their elongation at break properties after weathering for 3.5 $GJ.m^2$ in an aggressive climatic zone such as Townsville, Australia. Another example requires the retention of 50% impact resistance (notched impact) and color in accordance with specific LAB indices, after 6–24 months natural weathering or artificial weathering for 1000 hours (approximately one year of natural weathering).

In the case of thermal degradation, this process can be assessed by specific tests that compare mass loss before and after aging to assess the extent of the dehydrochlorination process. Dehydrochlorination can be measured by pH, color changes (LAB), changes in potential by chloride ion activity, and changes in conductance. One common method of assessing the level of residual stabilizer and the consequent resistance of PVC to dehydrochlorination, is the dehydrochlorination induction time (DIT). Stabilizers are incorporated into PVC resins to absorb free radicals and thus inhibit the degradation processes accelerated by these species, such as zip dehydrochlorination. DIT allows the measurement of the levels of stabilizer incorporated into the PVC resin. Stabilizer is used up during the extrusion process, and also by the effects of heat and UV radiation during external exposure, and consequently a certain residual level is required to allow adequate long-term performance. For window profiles it has been recommended that 50% of the initial level of stabilizer incorporated into the base resin should be left after processing (DIN 16830). While this is a sophisticated test and may be worthwhile considering for PVC pipe standards, modern PVC resins should be able to undergo extrusion and still have satisfactory levels of stabilizer. This test would only be considered necessary if USA manufacturers were to consider the inclusion of recyclate resin in their extrusion profiles, compounding of the resins themselves, or the sourcing of resins from countries that do not have a long history of PVC compounding.

Where there is potential for either UV or heat degradation, for example where PVC pipes are exposed or carry water at elevated temperatures (some southern areas of the USA), the PVC pressure pipe standards should consider incorporating suitable performance requirements to ensure long-term durability. While this report does not make specific suggestions on a representative test requirement, there are several suggestions that can be considered in the attempt to reduce the number of potential failures related to chemical degradation in PVC pipe applications, viz.:

- Protection from UV exposure and elevated temperatures during storage and transport should be specified in the appropriate installation codes, as for example in AWWA M23. While this current standard only specifies requirements for storage longer than two years, more rigorous requirements may be needed where exposure is in conditions that can be considered extreme, for example southern states of the USA.
- Where the pipe is stored outside for extended periods, the extent of degradation in storage could be assessed with an appropriate coloration and DIT, as discussed above.
- A minimum requirement for both UV and heat stabilizers could be specified, such that the expected rate of consumption of stabilizers will lead to a residual stabilizer level at the expected lifetime of the pipe. Oxygen induction times (OIT) and DIT could be used for such a process. An OIT specification is already common for polyethylene and ABS.
- Mechanical performance in the form of tensile properties, impact resistance, fracture toughness, and burst stress, could be assessed as a function of exposure to elevated service temperatures and weathering. The duration of exposure should be such that the PVC pipe is exposed to an equivalent dose of UV irradiation or elevated temperature. Artificial weathering and thermal exposures should only be performed when correlation to natural exposures has been proven, via physical assessments.
- Prescriptive requirements on the use of inorganic fillers such as TiO_2 could be made. A significant amount of work has been done on the levels required for specific applications, and one such study (Burn, 1992) recommends levels specific for PVC-U pipe applications.

MANUFACTURING COMPETENCY AND FIELD REQUIREMENTS

Generally, PVC is perceived to be a tough material that can be handled fairly roughly in the field. While this assumption may be incorrect and has led to a number of premature failures, it has also resulted in most international standards containing an impact strength requirement for the finished product (see Table 4.11). As discussed above, standards in the USA also have an impact requirement for the selected resin. This impact requirement (obtained using Izod impact) is defined in joules per meter of notch, and relates to the energy required to propagate a fast running crack in a notched Izod bar. These test specimens and the results obtained can be significantly removed from final performance, where the final impact resistance of the end product depends not only on resin selection, but also on processing levels/quality and pipe stiffness. Consequently, the definition of an impact requirement for the resin offers little advantage and may necessitate additional testing.

Most manufactures support an impact test because it is a simple test that can rapidly give an indication of poor manufacture, with the impact requirements defined in the standard as detailed in Table 4.12 being unrelated to field performance.

This would generally be the case, however for ASTM D2241, the test impact levels required have had a direct, though maybe unintended, effect. It can be seen from Table 4.12 that for pipes over 2 inches in size, the impact requirements for PVC-U are more than double those of the equivalent Australian standard, and for smaller pipe sizes, many times those of the ISO standard. It is known that pipes manufactured to the Australian standard using standard Ca/Zn-stabilized pipes occasionally fail at these impact levels, and significant unpublished work on Pb-stabilized pipes in this area using staircase impact assessment indicates that these pipes would have difficulty in consistently meeting the performance requirements of the USA standards.

Table 4.11
Required impact resistance (falling weight)

Standard	Minumum	Maximum
ANSI/AWWA C900-97	—	—
ANSI/AWWA C905-97	—	—
ANSI/AWWA C909-02	—	—
ASTM D1785-99	—	—
ASTM D2241	10 ft.lbf (0.25 inch)	160 ft.lbf (>8 inch)
ASTM F1483-98	100 ft.lbf (4 inch)	200 ft.lbf (12 inch)
AS/NZS 1477	7 ft.lbf (0.5 inch)	108 ft.lbf (>6 inch)
AS/NZS 4765 (Int)	1,441 ft.lbf (4 inch)	3,603 ft.lbf (>12 inch)
AS/NZS 4441 (Int)	43 ft.lbf (2 inch)	230 ft.lbf (>12 inch)
ISO 4422-1996	1.5 ft.lbf (0.5 inch)	92 ft.lbf (>12 inch)
ISO/DIS 16422	—	—

Table 4.12
Impact energies (ft.lbf) versus pipe size

Pipe size (inches)	ASTM 2241	ASTM 1483	AS 1477	AS 4765	AS 4441	ISO 4422
0.4	10	—	7.2	—	—	0.0
0.6	10	—	10.8	—	—	0.0
0.8	15	—	14.4	—	—	1.5
1	20	—	18.0	—	—	1.8
1.28	20	—	19.9	—	—	2.2
1.6	30	—	21.6	—	—	2.9
2	30	—	25.2	—	43.2	3.7
2.6	40	—	28.8	—	57.6	5.9
3.2	60	—	32.4	—	72.1	7.4
4	90	100	39.6	1,441.2	86.5	23.5
5	100	—	46.8	1,441.2	100.9	36.8
6	120	150	50.4	2,161.8	115.3	47.1
7	—	—	54.0	2,882.4	144.1	52.9
8	160	200	72.1	2,882.4	158.5	58.8
9	160	—	82.9	2,882.4	172.9	66.2
10	160	200	90.1	2,882.4	201.8	73.5
12	160	200	108.1	3,602.9	230.6	91.9
Test temperature (°C)	23	23	20	20	0	0

It was initially thought that this improved impact performance of pipes manufactured in the USA may have been due to the use of Sn stabilizers, which would lead to tougher polymers. The use of Sn stabilizers by themselves is known to lead to brittle polymers, and consequently correct lubrication is required to produce a satisfactory pipe. However, analysis of Australian and USA Sn-stabilized pipes has shown that the toughness can vary significantly, and that processing is a major factor in achieving the necessary toughness levels to adequately pass the impact test. Consequently, any review of PVC-U manufacturing standards in the USA will need to take the effect of these higher impact levels into account, because lowering them to those specified by ISO, for example, could have a possible effect on the failure rate experienced by water utilities in the future due to a potential increased number of failures due to slow crack growth.

Two standard manufacturing requirements that are not included in USA standards are a reversion and a high temperature requirement. These tests are common in many standards worldwide and help to ensure a consistent level of manufacture. The reversion test is carried out by immersing a pipe specimen in an atmosphere of 150°C (302°F) for a period of time, the specimen is then removed and the change in dimensions is recorded. A normal requirement is that the change in dimensions must not exceed 5%. By meeting this level, the manufacturer ensures that the levels of residual stress induced into the pipe by the haul-off process are minimized.

The high temperature test assesses the performance of both pipe and fittings at 180°C (356°F) and 150°C (302°F) respectively. After maintaining the specimen at the required temperature for one hour, it is sliced open and inspected for bubbles etc. The test has been shown to effectively assess pipes for bubbles resulting from incorrect processing conditions or thermal decomposition. The test also provides a qualitative indication of gelation levels in extruded PVC pipes. For fittings, it has been shown to be effective in detecting inclusions, voids, lack of fusion along weld lines, and residual stresses resulting from the extrusion molding process. In the Australian standard, the pipe component of this test originally had a reverse fold-back component, which was designed to detect poor spider-line fusion, however this test was found to be difficult to carry out for thicker walled pipes. This component of the test is being replaced with a flattening test, which is currently called up in USA standards, as detailed in Table 4.13.

Table 4.13
Flattening tests called up in PVC pressure pipe standards

Standard	Flattening percentage	Test period (minutes)
ANSI/AWWA C900-97	60	2–5
ANSI/AWWA C905-97	60	2–5
ANSI/AWWA C909-02	60	2–5
ASTM D1785-99	60	2–5
ASTM D2241	60	2–5
ASTM F1483-98	60	2–5
AS/NZS 1477		
AS/NZS 4765 (Int)		
AS/NZS 4441 (Int)		
ISO 4422-1996		
ISO/DIS 16422		

EFFECT ON WATER

Water quality issues are becoming a major concern to most water utilities. For example, PVC pipes manufactured in non-USA countries have come under increasing scrutiny from regulators because of the potential health effects from materials extracted from these types of water pipes. This is especially so for new distribution systems where the residence times for water can be large, and the levels of heavy metals such as Pb extracted can easily exceed the levels specified by WHO (Burn and Sullivan 1993). Because of uncertainty where pressure pipes can be used, some countries are requiring all manufactured pipes to be assessed as if they would be used for potable water applications. For example, Australian standards have recently required all pipes (regardless of their intended use) to pass a "materials in contact with drinking water" assessment because of these concerns.

Also, because of this concern over the potential health risks associated with the migration of substances from PVC pipes and other materials used in the water distribution system into potable water, all relevant USA standards for PVC pipe manufactured for use with potable water have a water quality requirement, as detailed in Table 4.14.

These water quality standards called up by the manufacturing standards include, among other quality requirements, the assessment of migration from products used in contact with potable water, and the regulation of the concentration for adverse substances. (Table 1.6 details the main standards available for the evaluation of materials in contact with drinking water.) In the case of ISO 16422, it only contains clauses that state that the materials shall not constitute a toxic hazard, shall not support microbial growth, and shall not give rise to any unpleasant taste or odor, cloudiness, or discoloration of the water.

Table 4.14
Water quality test requirements specified by PVC pipe standards

Standard	Water quality requirements	End product or compound
ANSI/AWWA C900-97	NSF 61	Both
ANSI/AWWA C905-97	NSF 61	Both
ANSI/AWWA C909-02	NSF 61	Both
ASTM D1785-99	NSF 14*	Pipe
ASTM D2241	NSF 14*	Pipe
ASTM F1483-98	NSF 61 or 14	Pipe
AS/NZS 1477	AS 3855 or AS 4020	Pipe
	VCM <1 mg/kg	Resin
AS/NZS 4765 (Int)	AS 3855 or AS 4020	Pipe
AS/NZS 4441 (Int)	AS 3855 or AS 4020	Pipe
ISO 4422-1996	ISO 3114 & <WHO levels	Pipe
ISO/DIS 16422	General statement only	

* The supplementary requirements of these standards require NSF 61 or the health effects portion of NSF 14 if the pipe is intended for transporting water.

In the USA, standards for pipes used to convey potable water require that the pipe or fitting is tested and certified according to ANSI/NSF 61, or as required by the health effect requirements in ANSI/NSF 14 for regulated substances. While in Australia, compliance with AS/NZS 4020 is required. European standards are not quite as prescriptive and only require that the "material shall have no adverse effects on the quality of drinking water", and they refer to the need for approval and regulations before the material can be introduced into the market (Europa 1998). Evaluation of such conditions can be performed using individual national standards, such as BS 6920, that require testing according to methods specified by the regulatory body, and to levels of contaminants specified by the regulator or by the WHO guidelines, whichever is the strictest. Alternatively, evaluation can be performed using ISO 8795, which is also a CEN standard.

ANSI/NSF 61 is currently a voluntary consensus standard available for public use. When referenced by federal, state, or local health authorities in their specifications, piping manufacturers must have their products evaluated against ANSI/NSF 61, and the registered trademark of the testing agency indicating certification should appear on the PVC piping products intended for use with potable water. As discussed in Chapter 1, ANSI/NSF 61 is very thorough in its requirement for health hazard evaluation, however, the extraction methodology could be simplified for time effectiveness without detriment to the hazard determination. It is recommended that the preconditioning period for the test sample in water be shortened from 14 days to 24 hours as in ISO 8795, if pipes are subjected to flushing periods during installation or, if more rigor is required, that the test sample be subjected to rinsing, as in AS 4030, followed by analysis of the first water sample collected after 24 hours to determine the most severe leachate rate. This would then be followed by the normal extraction procedure outline in ANSI/NSF 61. This recommendation is based on the observation that multiple flushing significantly reduces the levels of heavy metals available for leaching (Burn 1993).

INSTALLATION AND TRAINING

While manufacturing can play an important role in preventing premature pipeline failure, installation can play an even more important role for any pipe. For example, cases have been observed where the PVC pipe has been laid on point loads such as rocks and, in these cases, failure in periods as short as two years has been observed. Detailed installation requirements are provided by most reputable manufacturers, as well as by organizations such as AWWA, ASTM, and Uni-Bell. When pipe is installed according to these requirements, excellent performance can be obtained. This can be observed in Figure 4.2, where the failure rates for PVC pipe for Ottawa, Epcor and Calgary are relatively low. Epcor has taken major steps to ensure that PVC pipe is installed properly and, for a few years in the early 1990s, held a training program for all of its contractors that covered a range of issues such as jointing and bedding.

In the USA and Canada, national training programs do not exist in this area. National training programs do exist in countries such as Australia, where short-term technical training courses have been established for both PVC (Plastek) and PE installation (http://www.ntis.gov.au/, http://www.pipa.com.au/Training.html). These training programs are run by educational institutions, and some water utilities require installation contractors to have undergone training before they will be awarded a contract. It is recommended that suitable training courses be established in the USA, and that water utilities only use staff that have been adequately trained for pipe installation. An additional factor is supervision, and it has been observed in Australia that even where staff have been adequately trained, that without supervision, sloppy installation practices still occur. This can have a detrimental effect on the long-term performance of the pipe network, and all water utilities need to address the issue of training and supervision if they wish to ensure optimal performance from their water distribution network.

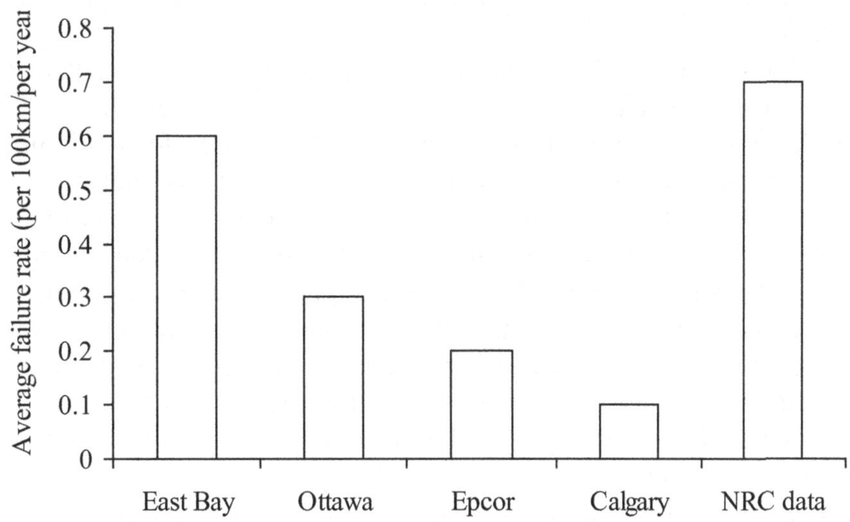

Figure 4.2 PVC-U failure rates for a number of Northern American cities

CONCLUSIONS AND RECOMMENDATIONS

Performance requirements for resin may be over-prescriptive, lead to excessive test requirements, limit manufacturer innovation, and will not guarantee a high-performance PVC pipe. It is suggested that resin selection be left to the pipe manufacturer, and that adequate performance requirements be included in pipe manufacturing standards.

The current method for calculating the HDB for PVC pressure pipes does not take data scatter into full account. Consideration should be given to adopting the ISO 9080 practice for HDB calculation, where the lower 97.5% confidence limit is used.

Design stresses within AWWA and ASTM standards are not consistent, and should be revised for consistency. The safety factors that define the final design stress should be based on a more scientific basis than those currently used. The current design stresses used within the USA for PVC-U and PVC-M are generally lower than those used elsewhere in the world, and require manufacturers to use more resin for similar performance.

The practice of adding a surge allowance for PVC pipe is questionable, and for most pipelines operating under static pressures this is not needed. For pipes operating under surge or fatigue conditions, the application of a surge allowance is simplistic and the pipeline should be specifically designed for these conditions.

Current short-term testing is prescriptive, does not confirm long-term pressure performance, and does not take resin variation into account. It is recommended that short-term pressure testing to confirm the long-term HDB be adopted. This testing should allow for changes in resin selection, and a suggested test procedure uses the 99.5% lower confidence limit value at 1 hour determined from the 50-year linear regression line for the appropriate resin, determined using ISO 9080.

Field failures invariably occur by slow crack growth followed by fast brittle fracture, rather than the ductile mechanism currently predicted by the standards. USA standards do not contain a performance requirement to minimize the propensity for PVC pipes to fail by a slow crack growth mechanism. Consideration should be given to include a fracture toughness performance requirement, as well as a test to ensure adequate gelation for PVC-U pipes, and a ductility test for PVC-M pipes. The current acetone test is not considered suitable for current PVC-U resins, as it is too benign and does not ensure adequate gelation levels. If a solvent test is still considered the most suitable method for determining gelation, the solvent should be changed to dichloromethane. Where occupational health and safety is an issue, another test method such as DSC should be considered, however no suitable test method has so far been developed, although an ISO study is in progress. The Australian industry is currently trialing an optical method that may be satisfactory for the USA industry.

Current USA standards do not specify any performance requirements for either thermal or UV stabilization. These factors may not be an issue for many parts of the USA, however, for southern climates where heat and UV may affect the longer term performance, consideration should be given to these two factors. While prescriptive factors such as defining a specific level of TiO_2 could be satisfactory, a better solution may be to define a minimum OIT and DIT. This would ensure that adequate stabilizers are incorporated into the PVC resin. An alternative would be to define minimum changes in pipe properties when the material has been exposed to a weathering regime.

The impact requirements of USA standards are significantly higher than those required elsewhere in the world. These impact requirements may have beneficial effects in that they require the PVC manufacturing processes used in the USA to produce more ductile pipes. However, this needs more investigation. It is recommended that the USA industry undertake a benchmarking study to determine the reasons for the higher ductility of Sn-stabilized pipes assessed in this study, and to determine if it is representative of standard manufacturing practices.

Both a high temperature oven test and a reversion test should be considered for inclusion into USA standards to ensure a high quality of manufacture.

Standards for ensuring the quality of materials for use in contact with drinking water are considered to be suitable to ensure that the pipe material does not contaminate potable water.

Consistency does not exist between AWWA and ASTM standards in the USA. A case exists for the development of one consistent standard for each of PVC-U, PVC-M, and PVC-O.

It is recommended that the industry in the USA develops national training courses for the jointing and installation of PVC pipes, and that attendance at these training courses become mandatory for any staff involved in the installation of PVC pipe by a water authority. Strategies for supervision of PVC pipe installation should also be developed by the industry, and these should be adopted as industry best practice.

CHAPTER 5
PERFORMANCE SURVEY OF PVC PIPES

In order to gauge the performance of PVC pipes in water supply networks, 44 water utilities in Australia, Canada, and the USA with significant volumes of PVC in their system were contacted to participate in a survey. Of those water utilities contacted, 17 water utilities provided detailed data, comprising 4 from the USA, 4 from Canada, and 9 from Australia. The survey requested general information regarding the total length of pipeline from each authority, and the proportion of PVC pipes in this length. More detailed information was requested on the annual failure rate (events/length/year) of PVC pipes in each water authority and, if available, a breakdown of the types of failures that occur in service. While this information allowed the current performance of PVC pipes to be compared for a number of water utilities, it provided raw data for benchmarking the failure models developed and discussed in Chapters 7 and 8.

Table 5.1 lists the water utilities that responded to the survey and the level of information that they were able to supply. As shown, of the 17 water utilities who answered the survey, all supplied details of the total length of their pipe network, and the percentage length of PVC pipes. Thirteen water utilities supplied estimates of the annual failure rate (per/length/per year) in their PVC pipes. Of these, 6 supplied a breakdown of the reported failure types that comprise this failure rate. Table 5.2 lists the total length of pipe networks, and percentage length of PVC pipes for each of the respondents.

Table 5.1
Current list of respondents to PVC pipe performance survey

Water authority	Information supplied		
	Length of PVC pipes in network	Annual failure rate	Breakdown of failure types
East Bay Municipal Utility District (EBMUD) (USA)	Yes	Yes	Yes
Des Moines Water Works (USA)	Yes	No	No
Louisville Water Company	Yes	Yes	Yes
City of Calgary Waterworks (Canada)	Yes	Yes	No
United Water Toms River (USA)	Yes	Yes	No
City of Richmond (Canada)	Yes	Yes	No
City of Ottawa (Canada)	Yes	Yes	Yes
Epcor (Canada)	Yes	Yes	No
City West Water (Australia)	Yes	Yes	Yes
Hunter Water (Australia)	Yes	Yes	Yes
South West Water Authority (Australia)	Yes	Yes	No
Yarra Valley Water (Australia)	Yes	Yes	Yes
South Australia Water (Australia)	Yes	No	No
Ipswich Water (Australia)	Yes	Yes	No
South East Water (Australia)	Yes	Yes	Yes
Gosford Water (Australia)	Yes	Yes	No
Sydney Water (Australia)	Yes	Yes	No

Table 5.2
Details of PVC pipes in water supply networks

Water authority	Total length of pipes in network (km)	Length of PVC pipe in network (km)	Percentage length of PVC pipes
EBMUD	6,417.6	513.4	8.0
Des Moines Water Works	1,546.6	241.2	15.6
United Water Toms River		336.1	
Louisville Water Company	5954.6	1012.3	17.0
City of Calgary Waterworks	4,146.0	1,699.9	41.0
City of Richmond (Canada)	633.8	157.9	24.9
City of Ottawa	2,577.7	515.5	20.4
Epcor	3,136.0	1098.5	35.0
City West Water	3,800.6	544.7	14.3
Hunter Water	4,558.4	626.1	7.3
South West Water	1,032.1	250.8	24.3
Yarra Valley Water	8,608.7	800.6	9.3
South Australia Water	24,489.0	587.7	2.4
Ipswich Water	1,318.7	250.6	19
South East Water	7,299.3	1,778.2	24
Gosford Water	938	24.4	2.6
Sydney Water	20,688	1,262.0	6.1

ESTIMATED ANNUAL FAILURE RATES IN PVC PIPES

Table 5.3 lists the estimated annual failure rate (failure/100km/year) for each water authority that supplied data. As shown, a wide range of failure rates (from 0.04 for Epcor to 6.8 for South East Water) were reported. It should be noted, however, that these are merely estimates, and were calculated by simply dividing the total number of recorded failures by the time period over which they were recorded and the *current* total length of PVC pipe. For example, Gosford Water recorded only three failures in the year 2001–2002 over a total PVC pipe length of 24.4 km. This corresponds to an estimated annual failure rate of 6.1 failures/100km/year.

In practice, the total length of PVC pipe and the number of failures will vary on a year-to year basis. Consequently, the annual failure rate will also change yearly. For a more accurate indication of annual failure rate (and how it changes with time), it is recommended that water utilities record and organize pipe asset and failure data as a minimum, in the format shown schematically in Table 5.4 (Jarrett et al. 2001) or more preferentially details of each failure including, time, date, failure type and location for each individual asset

By organizing pipe asset and failure data in this format, the changes in failure rates can be mapped accurately on a yearly basis. With sufficient failure data, this may also allow failure rates (length/year) to be extrapolated into the future. However, it should also be noted that for all water utilities, the volume of recorded failure data for PVC pipes failure is far less than that for other pipe materials. For example, Ipswich Water have recorded an estimated failure rate of 276 failures/100km/year for cast and ductile iron pipes, compared to 0.2 failures/100km/year for PVC pipes based on one failure. Consequently, in cases such as Ipswich Water, there is insufficient PVC pipe failure data for confident extrapolation and physical/probabilistic prediction curves need to be used for predicting future failures.

Table 5.3
Estimated annual failure rates for PVC pipes

Water authority	Length of PVC pipe (km)	No. of recorded failures	Recording period	Av failure rate (100km/y)	Av failure rate (100miles/y)
EBMUD	513.4	21	1996–2002	0.6	1.0
Des Moines Water Works*	241.263	0	N/A	0.0	0.0
United Water Toms River	336.1	8	2003–2004	1.7	2.7
Louisville Water Company	1,012.3	390	1976-2004	1.4	2.3
City of Calgary Waterworks	1,699.9	17	1991–2000	0.1	0.2
City of Richmond	157.9	4	2000-2004	0.5	0.8
City of Ottawa	515.5	16	1992–2002	0.3	0.5
Epcor	1,098.5	71	1977–2004	0.2	0.5
City West Water	544.7	206	1996–2003	5.4	8.7
Hunter Water	626.1	323	1992–2003	4.3	6.9
South West Water	250.8	Unknown	Unknown	1.5	2.4
Yarra Valley Water	800.6	384	1995–2003	5.3	8.5
South Australia Water	587.7	Unknown	Unknown	Unknown	Unknown
Ipswich Water	250.6	1	2000–2001	0.2	0.3
South East Water	1,778.2	844	1996–2002	6.8	10.9
Gosford Water	24.4	3	2000–2001	6.1	9.8
Sydney Water	1,262.0	134	1997–2002	1.8	2.9

* Does not include failures for joints and/or external damage.

Table 5.4
Pipe asset and failure data organization

Asset no.	Year laid	Age (years)									Length
		0	1	2	…	10	11	…	…	40	(m)
001	1980	X*	X	X	X	X	1	2	0	2	101
002	1981	X	X	X	X	2	0	0	0	1	203
Etc.	1983	X	X	…	…						

* "X" indicates that no data were recorded.

BREAKDOWN OF FAILURE TYPES IN PVC PIPES

For those utilities that supplied a breakdown of pipe failures, the following descriptions were requested.

Blown Section

This corresponds to a pipe failure when a section of the pipe wall has been removed. In practice, this can occur when a propagating brittle fracture bifurcates (i.e., a second crack initiates) and then rejoins the original crack path. The resulting bounded section of the pipe wall is then free, as shown schematically in Figure 5.1.

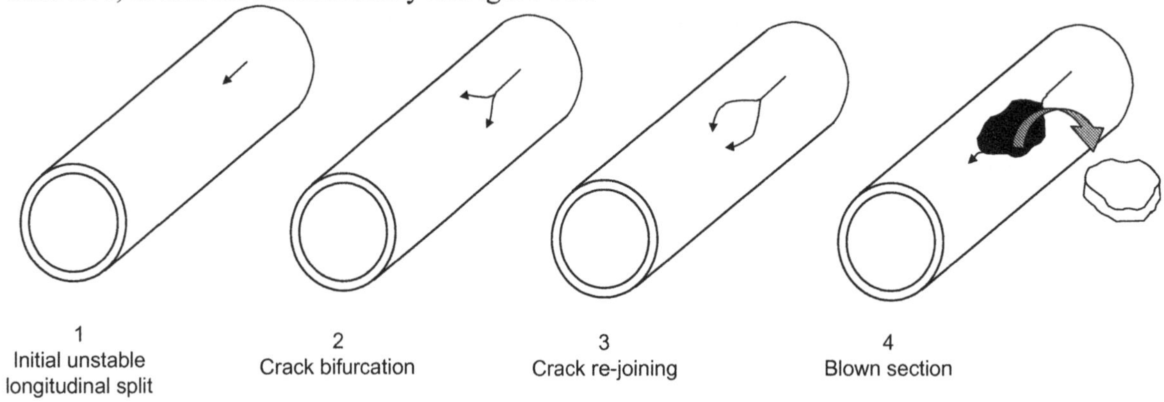

Figure 5.1 Schematic illustration of how a blown section can occur in PVC pipe

Circumferential Fracture

This corresponds to a single crack initiating and propagating around the circumference of the pipe. Since the direction of crack propagation is perpendicular to the direction of maximum principal stress, this failure mode suggests that the pipe has experienced axial stresses, which can originate from a number of potential sources. For example, lateral ground movement or soil expansion around a poorly supported pipe can generate axial bending stresses. Alternatively, friction at the pipe/soil interface can restrain axial thermal expansion experienced upon changes in ground temperature. In turn, this can induce direct axial tensile stress.

Longitudinal Fracture

This failure mode corresponds to a single crack initiating and propagating axially along the length of the pipe. These failures can be either short or long in length. However, from the data collected, the length could not be differentiated. The length of the fracture can offer significant information on the quality of manufacture, or on the operating conditions the pipe was installed under.

Tapping Band

This corresponds to failure associated with tapping bands. This may be a failure in the tapping band itself, or a pipe failure that occurs during tapping.

Other

This corresponds to any failure mode that cannot be classified by the above descriptions. Where available, water utilities were requested to use their in-house descriptions for these failure modes. The distribution of failure modes for the relevant water utilities is plotted in Figures 5.2–5.8.

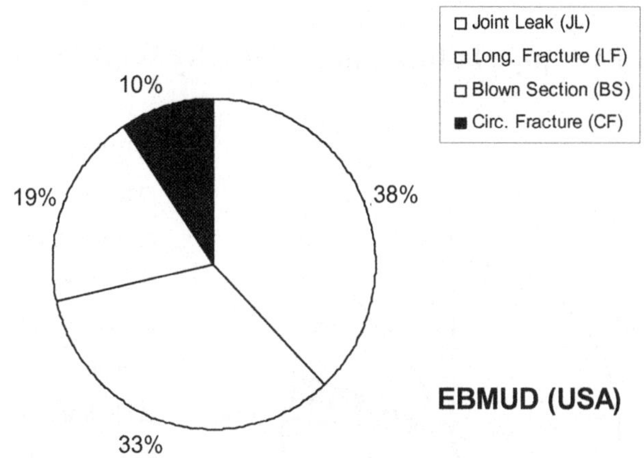

Figure 5.2 Breakdown of PVC pipe failure modes for EBMUD (USA)

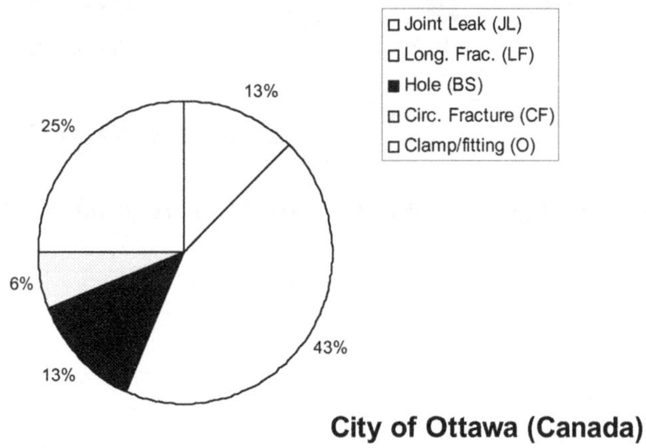

Figure 5.3 Breakdown of PVC pipe failure modes for City of Ottawa (Canada)

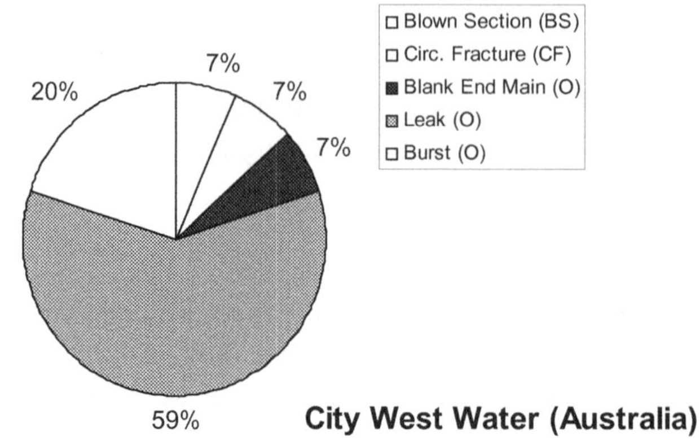

Figure 5.4 Breakdown of PVC pipe failure modes for City West Water (Australia)

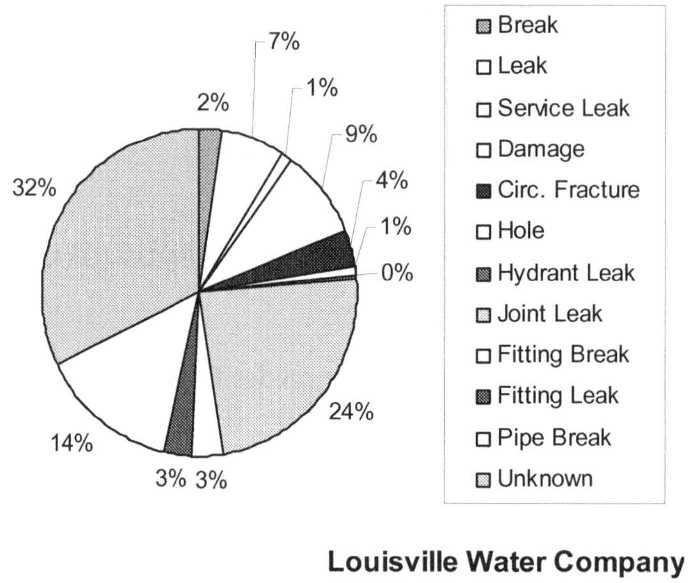

Figure 5.5 Breakdown of PVC pipe failure modes for Louisville Water Company (USA)

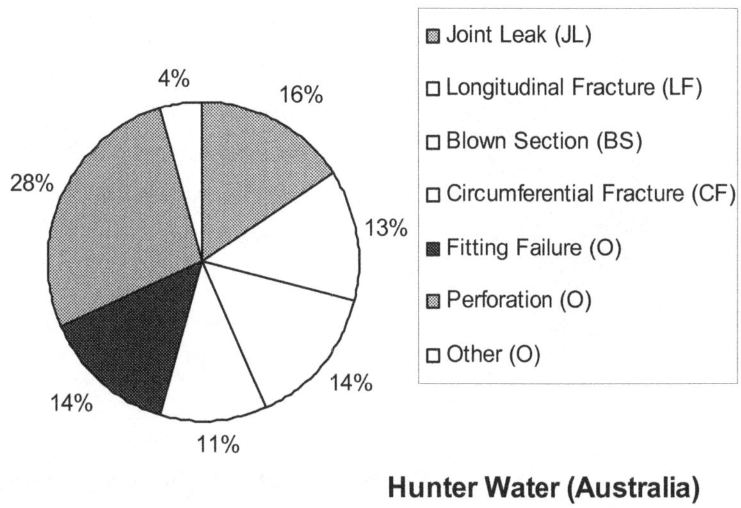

Figure 5.6 Breakdown of PVC pipe failure modes for Hunter Water (Australia)

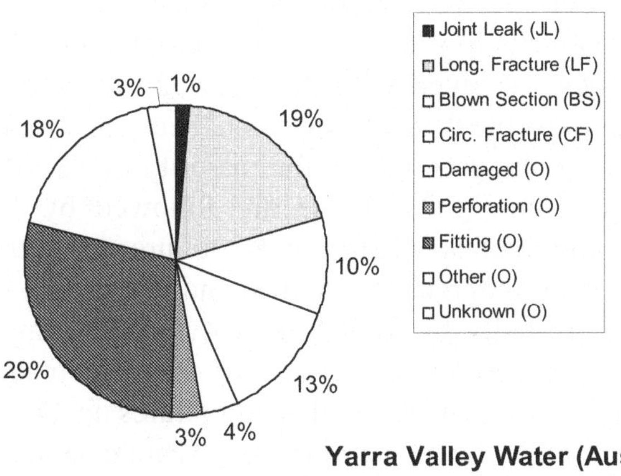

Figure 5.7 Breakdown of PVC pipe failure modes for Yarra Valley Water (Australia)

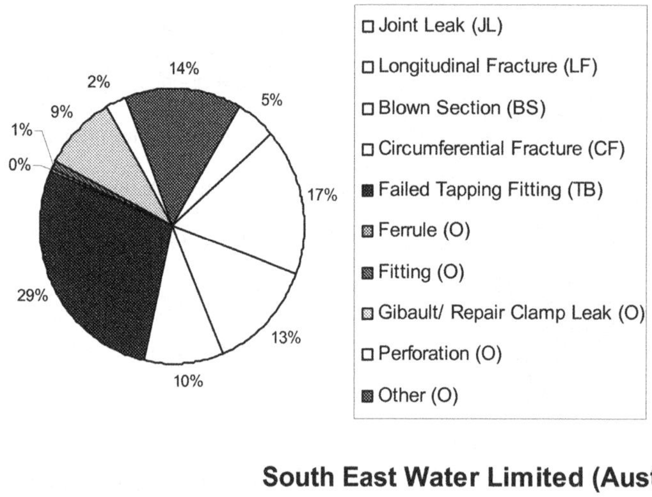

Figure 5.8 Breakdown of PVC pipe failure modes for South East Water (Australia)

As shown in Figure 5.2, joint leak failures account for the highest percentage of PVC failures for EBMUD. These are followed by longitudinal fractures, blown sections, and finally circumferential fractures. The relatively low numbers of circumferential fractures is intuitive since the flexibility of plastic pipes combined with the allowed movement of a rubber-ringed joint should accommodate ground movement and reduce axial bending stresses.

In contrast, longitudinal fractures account for the majority (43%) of recorded PVC pipe failures for the City of Ottawa (Figure 5.3). These are followed by failures classified as "clamp/fitting" (25%). It is currently unclear whether these failures are attributed to fracture or leakage from fittings and clamps. Circumferential fractures and blown sections each account for 13% of recorded failures, followed finally by joint leaks (6%). While the previous arguments regarding pipe flexibility apply here, the slight increase in the percentage of circumferential fractures may be attributed to relatively low ground temperatures in Ottawa. As outlined by Rajani et al. (2000), frictional restraint at the pipe/soil interface can generate axial tensile stresses as the pipe "attempts" to contract at low temperatures. Although axial contraction can be accommodated by the rubber ring joints in PVC pipes, frictional restraint between the pipe outer surface and the surrounding soil can be sufficient to generate axial tensile stresses (Rajani et al. 2000).

Following a similar argument, the relatively high percentage of longitudinal fractures may be attributed to frost loading and diametrical deflection of pipes. This would increase the circumferential stresses already generated by internal pressure. However, anecdotal evidence also suggests that in Ottawa, PVC pipes are installed to a depth of 2.4 metres, which is thought to be effectively below the frost line (Holmes, 2004). Longitudinal fractures in the City of Ottawa may also be attributed to poor installation practice. For example, over insertion of pipes into rubber ring joints can generate excessive circumferential tensile stress resulting in longitudinal fracture propagating along the pipe from the socket (Holmes, 2004). An additional failure mechanism observed in the City of Ottawa is erosion of the pipe wall thickness in the vicinity of a leaking joint. As a joint leaks and washes sand backfill over the pipe outer surface, local erosion occurs, which reduces the pipe wall thickness until it cannot resist the applied stress and a puncture occurs (Holmes, 2004). Although this complex failure mechanism differs from longitudinal fracture, it is likely that these failures would have been reported using this description. It should be noted that PVC pipelines in Ottawa use push-on joints for pipe to pipe joints and with small (100 to 200mm) AWWA C907 PVC injection molded fittings. Ottawa however uses mostly mechanical joints for all its Ductile Iron fittings, gate valves and hydrants. Mechanical joints are possibly more prone to leaks through improper tightening than simpler push-on joints which reduce the possibility of human error from incorrect joint tightening. Perhaps this is contributing to the lower break rates enjoyed by some municipalities such as Edmonton/Epcor. (Holmes, 2004). This may be an indirect cause of longitudinal fractures through the erosion mechanism proposed above. Anecdotal evidence exists to suggest that longitudinal fractures are observed less frequently in PVC pipelines that use simpler push-in joints on fittings that are perhaps less likely to be installed improperly (Holmes, 2004).

For City West Water (Figure 5.4), the majority of PVC pipe failures are recorded as leaks (59%). It is unclear whether this refers to a pipe or joint failure. The definition of burst is also unclear. For Louisville Water Company (Figure 5.5), the pipes are broken up in to twelve failure categories, with the majority of failures attributed to an unknown cause (32%). These were further simplified into breaks (57% of all failures) and leaks (43% of all failures), the unknown failures were attributed to the breaks category.

An examination of the failure breakdown for Hunter Water (Figure 5.6) indicates that the recorded failures are spread mainly between perforation (28%), blown section (14%), fitting failure (14%), joint leak (16%), longitudinal split (13%), and circumferential failure (11%), with the rest of failures (4%) classified as other. The relatively low percentage of circumferential fractures is attributed to the flexibility of PVC compared to rigid pipe materials, and the fact that lateral pipe movement can be largely accommodated by rubber-ring spigot/socket joints. In comparison to Ottawa (Figure 5.3), ground temperatures in Australia are not low enough to induce circumferential fractures from the frictional constraint mechanism discussed above.

As shown in Figure 5.7, the most comprehensive breakdown of failure modes was supplied by Yarra Valley Water, where 29% of recorded failures were attributed to fitting failures. Again, further information is required to determine whether this corresponds to fracture or leakage from fittings. Nineteen percent of recorded failures are attributed to longitudinal fractures, followed by 18% which are classified as other. It is currently unclear what failure mode this classification indicates. Thirteen percent of failures are classified as circumferential fractures (which is relatively high compared to the other Australian water utilities), and 10% of failures are attributed to blown sections. It is interesting to note that only 4% of recorded failures are classified as damaged. While further information is required, this suggests that, in this case, only a small percentage of PVC pipe failures are caused by external damage, such as that incurred by a third party. It also interesting to note that 3% of pipe failures are classified as perforations, which is a description normally reserved for through-thickness corrosion pits in metallic pipes. In practice, perforations in PVC pipes are probably very small longitudinal splits created by slow crack growth through the pipe wall. If the applied SIF (see Chapter 3) does not exceed the material fracture toughness *before* the crack traverses the full wall thickness, then such a split will result. This is also referred to as a "leak before break" failure (Lu, Davis and Burn. 2003). The remaining 1% of PVC pipe failures for Yarra Valley Water are classified as joint leaks.

For South East Water, as with Yarra Valley Water, a large proportion of failures are attributed to tapping fittings (Figure 5.8). Again, it is unclear whether this corresponds to pipe failure underneath the fitting, or leakage from the fitting body itself. It is interesting to note the proportions of circumferential and longitudinal fractures are similar to Yarra Valley Water. The slightly lower proportion of circumferential fractures for South East Water (9.9% compared to 13% for Yarra Valley Water) may reflect the smaller area of expansive clay soils that is known to exist in the South East Water network area. This would lead to fewer instances of lateral soil movement creating bending and axial tensile stresses in poorly supported pipes.

FAILURE RATES IN CANADIAN PVC PIPES

For comparison with the current field performance survey, previously recorded failure data for PVC pipes in Canadian water utilities was made available in a study conducted by the National Research Council (NRC) of Canada (Rajani, McDonald and Felio 1995). Data from 21 Canadian water utilities were used to determine average failure rates and different failure modes in PVC pipes. The percentage length of PVC pipes for each water authority in this study is listed in Table 5.5.

Where available, each water authority in the NRC study supplied failure data for PVC pipes recorded in 1992 and 1993. The estimated average failure rates (failures/100km/year) from these data are given in Table 5.6.

As shown, the majority of water utilities with PVC pipes have not observed failures. However, it should be noted that compared to the Australian water utilities, the total length of PVC pipes in many Canadian utilities is relatively small. Summing all PVC failures over the total PVC pipe length in Canadian water utilities gives an average failure rate of 0.7 failures/100km/year (Rajani, McDonald and Felio 1995). Referring back to Table 5.3, a significantly higher average failure rate of 4.6 failures/100km/year is obtained if Australian PVC failure data is aggregated in a similar manner. The breakdown of failure modes for aggregated failures in Canadian PVC pipe is shown in Figure 5.9.

Compared to Australian utilities, a larger fraction of total failures in PVC pipes occurs by longitudinal fracture in Canadian water utilities (47% in 1992 and 22%1993). One possible reason for this difference is the action of frost loading on Canadian pipes. It is possible that this additional load will increase the in-plane bending stress on a buried pipe, which contributes to longitudinal fracture. However, comparisons should be interpreted carefully, since significantly fewer PVC failures have been observed in Canadian authorities and reported over a relatively short time period

Table 5.5
Percentage of PVC pipes in Canadian water utilities

City/region	Total length of pipes in network (km)	Length of PVC pipe in network (km)	Percentage length of PVC pipes
Victoria	390.5	32.6	8.3
Vancouver	1,383.0	0.0	0.0
Burnaby	635.5	6.6	0.0
Calgary	3,322.0	783.0	23.6
Edmonton	2,628.4	444.4	16.9
Regina	713.3	147.9	20.7
Moose Jaw	256.4	25.9	10.1
Peterborough	357.3	22.7	6.3
Pembroke	91.9	10.1	10.9
Ottawa–Carleton	2,300.0	46.0	2.0
North York	1,352.4	82.8	6.1
Windsor	888.3	43.2	4.9
London	1,257.3	63.5	5.1
Sherbrooke	373.0	85.8	23.0
Fredericton	287.1	2.9	1.0
Halifax	445.0	0.5	0.1
Sydney	144.7	0.0	0.0
Summerside	75.1	18.8	25.0
St John's	412.4	0.0	0.0
Corner Brook	140.0	1.4	1.0
Mount Pearl	100.0	0.0	0.0

Table 5.6
Estimated failure rates in PVC pipes in Canadian water utilities

City/region	Total length of PVC pipes in network (km)	Failures per 100km in 1992	Failures per 100km in 1993	Average failure rate (failures/100km/year)
Victoria	32.6	0.0	0.0	0.0
Vancouver	0.0	0.0	0.0	0.0
Burnaby	6.6	0.0	0.0	0.0
Calgary	783.0	0.0	0.0	0.0
Edmonton	444.4	0.0	0.0	0.0
Regina	147.9	0.0	0.0	0.0
Moose Jaw	25.9	0.0	0.0	0.0
Peterborough	22.7	0.0	0.0	0.0
Pembroke	10.1	0.0	0.0	0.0
Ottawa–Carleton	46.0	6.0	0.0	3.0
North York	82.8	1.7	0.8	1.2
Windsor	43.2	0.0	1.7	0.9
London	63.5	3.4	1.7	2.6
Sherbrooke	85.8	1.7	0.0	0.9
Fredericton	2.9	0.0	0.0	0.0
Halifax	0.5	0.0	0.0	0.0
Sydney	0.0	0.0	0.0	0.0
Summerside	18.8	5.1	0.0	2.6
St John's	0.0	0.0	0.0.	0.0
Corner Brook	1.4	0.0	0.0	0.0
Mount Pearl	0.0	0.0	0.0	0.0

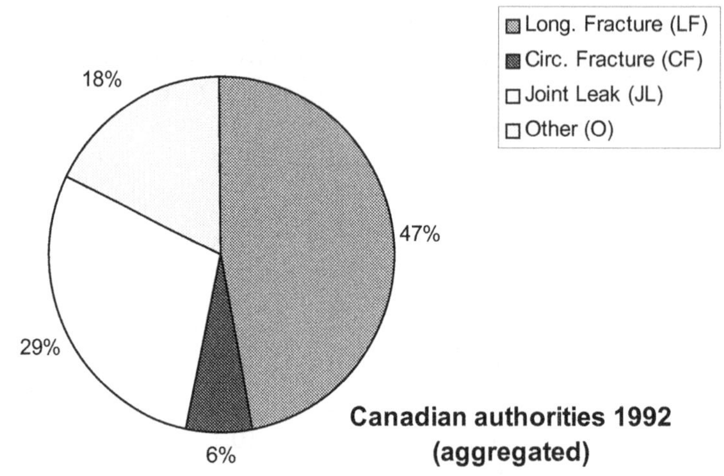

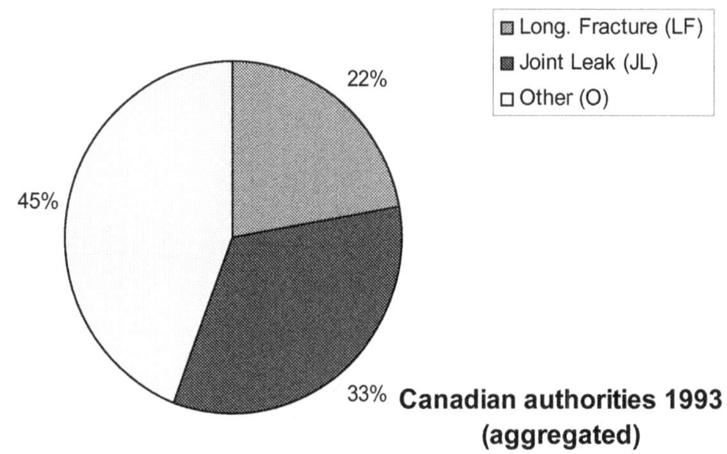

Figure 5.9 Breakdown of PVC pipe failure modes for Canadian water utilities in 1992 and 1993

FAILURE RATES IN UK PVC PIPES – UKWIR DATA

For comparison with the current field performance survey, previously recorded failure data for PVC pipes in UK water utilities was made available by UK Water Industry Research (UKWIR).

In a study conducted by MacKellar and Pearson (2003), a national database of water mains failures was compiled from detailed pipe asset and failure data provided by 17 water utilities across the UK. A summary of recorded average failure rates in different pipe materials in the UK database between 1998 and 2002 is given in Table 5.7

As shown, PVC pipes in the UK exhibit average failure rates between 7.2 and 9.6 failures/100km/per year between the recording years 1998 to 2001. In comparison to Tables 5.3 and 5.6, this range in average failure rate is comparable (if slightly higher) to that reported by Australian water utilities, but significantly higher than the range reported by USA/Canadian utilities. Figure 5.10 gives a breakdown of failure modes in PVC pipes recorded by UK water utilities.

Table 5.7
Estimated failure rates in different pipe materials in UK water utilities (Data from MacKellar and Pearson, 2003)

Average failure rate (failures/100km/per year)	Year					
Material Group	1998	1999	2000	2001	2002[1]	Grand total
Asbestos Cement	16.4	17.1	15.1	15.8	15.6	16.0
Ductile Iron	5.0	5.3	4.8	4.8	6.5	5.3
Cast Iron	23.7	23.7	19.1	21.7	12.3	20.1
Polyethylene	3.5	2.9	3.3	3.1	3.0	3.2
PVC	**9.6**	**9.1**	**7.2**	**7.4**	**3.3**	**7.3**
Steel	5.0	6.1	5.8	5.7	33.1	11.1
Unknown	0.1	0.0	0.0	0.1	15.9	3.2
Grand total	9.0	9.2	7.9	8.4	12.8	9.5

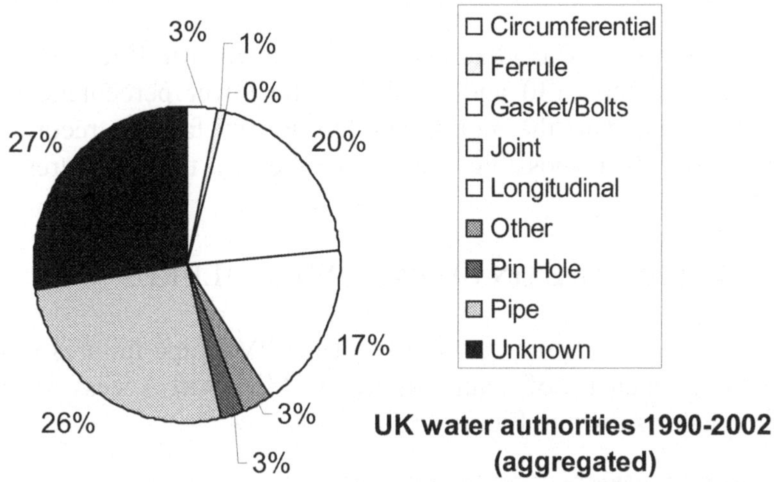

Figure 5.10 Breakdown of PVC pipe failure modes for UK water utilities between 1990 and 2002

As shown, 27% of recorded failures are classified as "unknown". This is followed by 26% classified as "pipe". It is unclear which particular PVC pipe failure mode this description refers to 20% of recorded failures are described as "joint" which probably corresponds to leaking joints, and 17% of failures are classified as "longitudinal" corresponding to longitudinal fracture of a pipe. While 3% of failures are classified as "Pin Hole", this probably corresponds to very short longitudinal fractures, created by slow crack growth through the pipe wall (a similar failure mode "perforation" was reported by Yarra Valley Water (Australia)). Only 3% of recorded

[1] The 2002 figures are unreliable since this was an incomplete recording year. The steel figures are particularly skewed by a single water authority (MacKellar and Pearson, 2003)

failures are classified as "circumferential" corresponding to circumferential fracture. As discussed previously, this may reflect the extent of lateral pipe movement afforded by rubber ring joints in PVC pipes, which allow significant deflections to take place in poorly bedded pipes, without generating any axial bending stress.

Table 5.8 compares dominant failure modes for PVC pipes in USA/Canadian, Australian and UK water utilities. Three categories are listed; Circumferential Fracture, Joints/Fittings (comprising all failures recoded as "joint leaks" or "fitting failures") and Longitudinal Fracture (this also includes failures reported as "blown section", "pin hole" and "perforation").

Table 5.8
Comparison between PVC failure modes in USA/Canadian, Australian and UK water utilities

Water utilities	Circumferential Fracture (%)	Joints/Fittings (%)	Longitudinal Fracture (%)
USA/Canada	5.4	34.3	36.3
Australian	10.2	40.5	30.0
UK	3.0	21.0	20.0

As shown, both USA/Canadian and Australian water utilities exhibit a similar fraction of longitudinal fracture failures (30 and 36.3%). The lower percentage reported by UK water utilities may be attributed to the significant fraction of failures recorded as "unknown" and "pipe" (Figure 5.9). It is possible that a number of these failures actually occurred by longitudinal fracture.

THE EFFECT OF PIPE AGE ON UK PVC PIPE FAILURES

In addition to a report summarising recorded PVC pipe failures in the UK, UKWIR also supplied the national database of mains failures in Microsoft Access form. An example of the type of data included is shown in Figures 5.11 and 5.12.

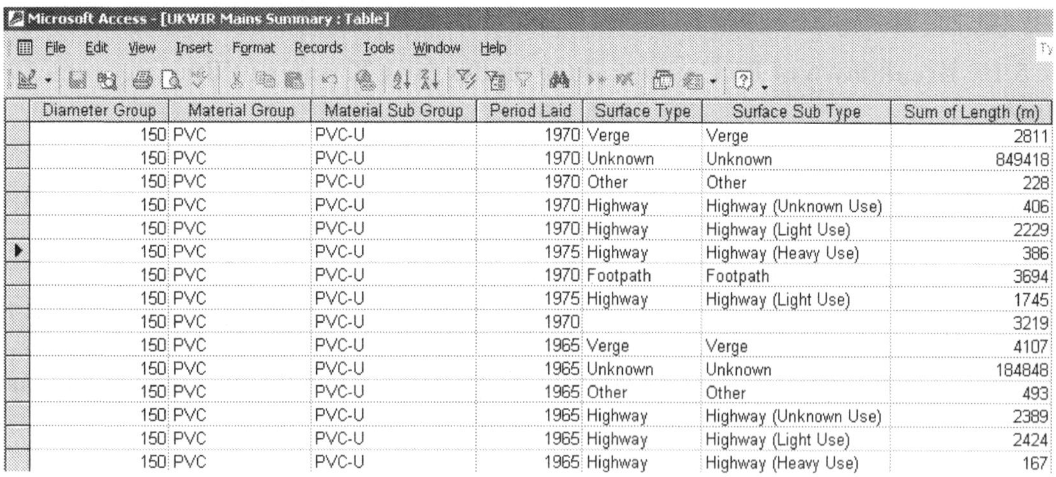

Diameter Group	Material Group	Material Sub Group	Period Laid	Surface Type	Surface Sub Type	Sum of Length (m)
150	PVC	PVC-U	1970	Verge	Verge	2811
150	PVC	PVC-U	1970	Unknown	Unknown	849418
150	PVC	PVC-U	1970	Other	Other	228
150	PVC	PVC-U	1970	Highway	Highway (Unknown Use)	406
150	PVC	PVC-U	1970	Highway	Highway (Light Use)	2229
150	PVC	PVC-U	1975	Highway	Highway (Heavy Use)	386
150	PVC	PVC-U	1970	Footpath	Footpath	3694
150	PVC	PVC-U	1975	Highway	Highway (Light Use)	1745
150	PVC	PVC-U	1970			3219
150	PVC	PVC-U	1965	Verge	Verge	4107
150	PVC	PVC-U	1965	Unknown	Unknown	184848
150	PVC	PVC-U	1965	Other	Other	493
150	PVC	PVC-U	1965	Highway	Highway (Unknown Use)	2389
150	PVC	PVC-U	1965	Highway	Highway (Light Use)	2424
150	PVC	PVC-U	1965	Highway	Highway (Heavy Use)	167

Figure 5.11 Example of UKWIR national database: Mains Summary

Burst Type	Burst Sub Type	Burst Month	Burst Year	Diameter Group	Material Group	Material Sub Group	Period Laid	Surface Type	Surface Sub Typ
Unknown	Unknown	9	1997	100	PVC	PVC-U	1995	Unknown	Unknown
Joint	Joint	6	1999	100	PVC	PVC-U	1975	Unknown	Unknown
Pipe	Circumferential	10	1999	100	PVC	PVC-U	1995	Unknown	Unknown
Unknown	Unknown	10	1998	100	PVC	PVC-U	1960	Highway	Highway (Unknow
Unknown	Unknown	9	1997	100	PVC	PVC-U	1960	Footpath	Footpath
Pipe	Longitudinal	2	1999	100	PVC	PVC-U	1940	Unknown	Unknown
Unknown	Unknown	9	1997	100	PVC	PVC-U	1965	Unknown	Unknown
Pipe	Circumferential	10	1999	100	PVC	PVC-U	1995	Unknown	Unknown
Unknown	Unknown	12	1997	100	PVC	PVC-U	1960	Unknown	Unknown
Pipe	Longitudinal	10	1998	100	PVC	PVC-U	1975	Unknown	Unknown
Pipe	Circumferential	6	1999	100	PVC	PVC-U	1975	Unknown	Unknown
Unknown	Unknown	8	1997	100	PVC	PVC-U	1970	Verge	Verge
Unknown	Unknown	9	1997	100	PVC	PVC-U	1985	Unknown	Unknown
Pipe	Circumferential	10	1998	100	PVC	PVC-U	1995	Unknown	Unknown
Joint	Joint	10	2000	100	PVC	PVC-U	1940	Unknown	Unknown
Unknown	Unknown	9	1997	100	PVC	PVC-U	1965	Footpath	Footpath

Figure 5.12 Example of UKWIR national database: Burst Summary

As shown, the UKWIR database is organised into a "Mains Summary" and a "Burst Summary". The Mains summary contains information describing the diameter, material and installation year for different materials. The type of surface and length of main installed is also specified. The Burst Summary contains information describing individual failures in pipes. A shown in Figure 5.12, the type of failure is given along with the month and year that the failure was recorded. The pipe diameter, material and installation year is also supplied.

Combining the information on the Mains and Burst summaries, the effect of age on average failure rates for UK PVC pipes can be determined. An average failure rate (failures/pipe length/year) can be defined for a particular pipe age as (Jarrett et al. 2001):

$$Average\ failure\ rate\ at\ age\ A = \frac{Number\ of\ observed\ failures\ at\ age\ A}{Exposed\ pipe\ length\ at\ age\ A} \tag{5.1}$$

For a particular pipe age, the exposed pipe length is that length of pipe that is the same age at the time of failure (Jarrett et al. 2001). An algorithm to calculate average failure rates for pipes of different ages in the UKWIR database has been written using the Python 2.3 programming language. The algorithm is listed in Appendix C, and the resulting graph of average failure rate (failures/100km/per year) vs. age for PVC-U pipes in the UKWIR database is given in Figure 5.13

As shown, average failure rate increases at a very low rate with pipe age (from 1 failure/100/km/per year, to 4 failures/100km/per year) until an increase occurs for pipes greater than 35 years old. The 95% confidence limits associated with each estimate of average failure rate is approximately given by (Jarrett et al. 2001)

$$\text{Upper limit} = \text{Av. Failure Rate} \times \left[\exp\left(\frac{2}{\sqrt{n}}\right)\right] \tag{5.2}$$

$$\text{Lower limit} = \text{Av. Failure Rate} \div \left[\exp\left(\frac{2}{\sqrt{n}}\right)\right] \tag{5.3}$$

Where n is the number of observed failures that comprise the estimated average failure rates. Since the installed length (and hence exposed length) of these older pipes is relatively low, fewer failure events have occurred and the error associated with these 95% confidence limits average increases. However, the observed increase in average failure rate falls within the limits set by this error, suggesting that the effect of pipe age is valid.

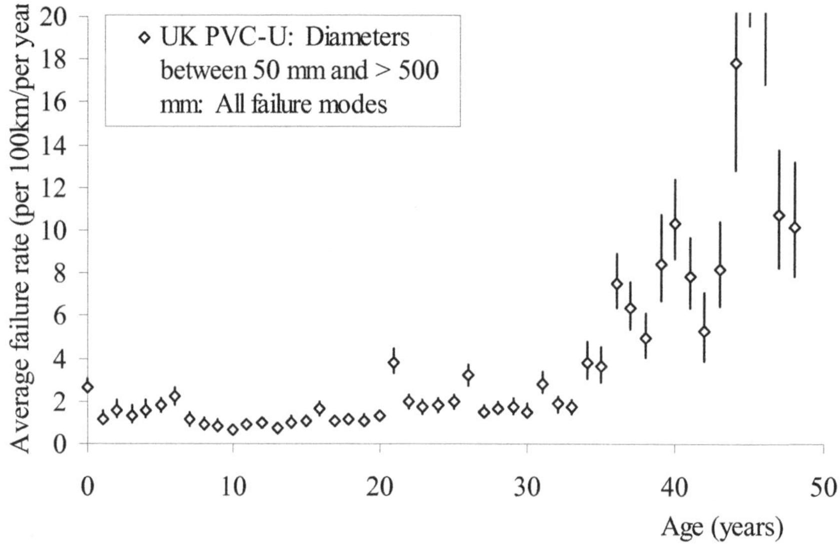

Figure 5.13 Average failure rate vs. age for UK PVC-U pipes: All failure modes in diameters between 50 mm and > 500 mm

To assess the influence of failure mode in this observed influence of pipe age, Figures 5.14 and 5.15 show average failure rates obtained from fracture failures only (i.e. longitudinal, circumferential and "pin hole" fractures) and joint/fittings failures only.

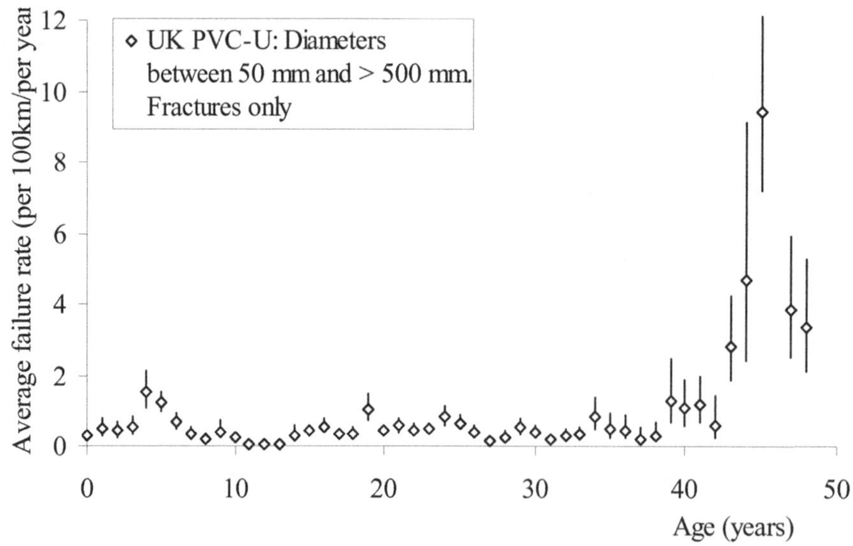

Figure 5.14 Average failure rate vs. age for UK PVC-U pipes: Fracture failure modes only in diameters between 50 mm and > 500 mm

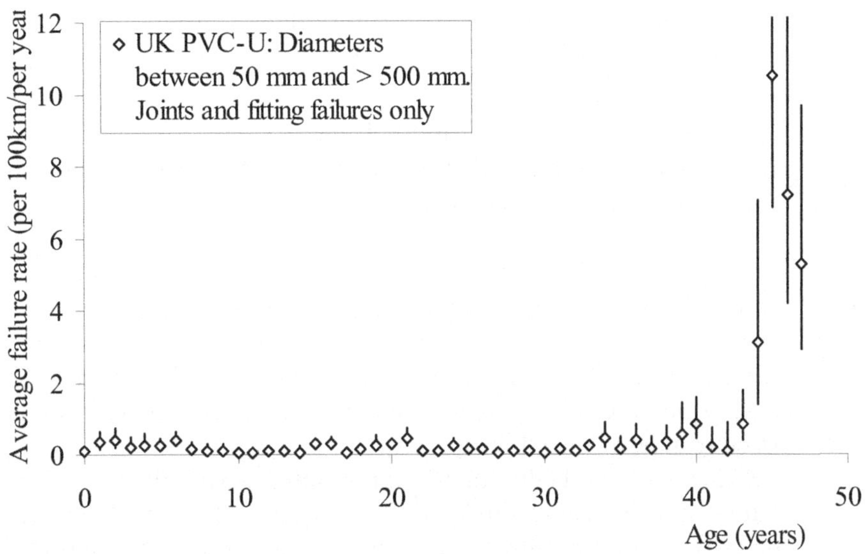

Figure 5.15 Average failure rate vs. age for UK PVC-U pipes: Joint and fitting failure modes only in diameters between 50 mm and > 500 mm

As shown, in both cases, a relatively constant average failure rate is observed until a sharp increase at a pipe age of around 42 years old. A sharp increase is then observed as average failure rates to around 10 failures/100km/per year for pipe that are around 45 years old. Again, since the exposed length of these older pipes is lower, and fewer failures have been observed, the error associated with estimates of average failure rate increases. However, since the increase is within the error set by the 95% confidence limits, it is assumed to be valid.

It is possible that the sharp increase in fracture failure rate in Figure 5.14 is related to changes in PVC pipe production techniques and/or installation techniques. For example, according to Alferink, Janson and Holloway (1997), a change from granulated to powdered PVC was made in Europe in 1971. The result of this change was to improve gelation in extruded PVC pipes and to increase fracture toughness levels. Consequently, as noted by Marshall, Pearson and MacKellar (1996), fracture failures in PVC pipes installed before this period were "largely restricted to the UK and Holland where poor control of pipe processing had led to low static toughness levels". For pipes installed after 1971, processing control and fracture toughness levels had improved, which would reduce the number of fracture failures observed in the field. Additionally, the improvement could be attributed to the adoption of the first UK Code of Practice which was introduced in 1973, addressing transport, handling, and installation of plastics pipes (BSI 1973). The introduction of this code may explain the abrupt reduction in joint/fitting failures as observed in Figure 5.15.

STANDARD FAILURE CLASSIFICATIONS

The survey carried out in this study shows that there is considerable variation in the way that water utilities report their failure data for PVC pipes. A standardized methodology would make comparison of failure information between water utilities much easier. Recently the Water Services Association of Australia (WSAA) has recommended that a standardized reporting methodology be used by its members for all pipes types, and the adoption of a similar process by water utilities in the USA would make analysis of failure data to predict long-term failure trends much easier. This methodology is defined in Table 5.9.

Table 5.9
WSAA terminology for reporting failures in water distribution pipes

CSIRO/WSAA designation	Failure mode to be grouped under this description
Blown section	Burst, burst main, burst section, broken section, blown section
Circumferential fracture	Broken back, circumferential fracture, circumferential split
Ferrule	Ferrule replacement, saddle and ferrule, other (specify) – ferrule, ferrule out of main, split in ferrule, leaking ferrule, ferrule out, leak on ferrule, supply court, fracture at ferrule, split ferrule
Fitting failure	Leaking washout, faulty fittings, fitting split, leaking fitting, leak on tee, fitting failure, hole in bend, corroded fitting, washout, blank end blow off, replace defective T, broken fitting, faulty fitting, blow out on bend, blown P-bend
Gibault/repair clamp leak	Bolts on gibault corroded, gibault/repair clamp leak, leak on clamp, blown gibault rubber, bolts on gibault corroded, leaking gibault joint, broken fitting gibault coupling, corroded bolts, faulty SS clamp
Joint leak	Blown rubber on flange, blown joint, lead joint, joint leak, leak on collar, leaking rubber ring
Longitudinal split	Split on bend (PVC), split in main, split pipe, longitudinal split, longitudinal fracture, spiral fracture, split, split main
Other	Blank ends, no information, job not finished, blown T, other, damaged, not stated, poor supply, fire plug riser split, ball valve, end blown off
Perforation	Perforation, pinhole in pipe (note that PVC and PE perforations may be splits)
Failed tapping fitting	Tapping band, leaking saddle, saddle, tapping section, tapping saddle, broken band, blown saddle, blown tapping saddle, broken saddle
Unknown	No type entered

A number of failed pipe samples have been received from Australian and Canadian water utilities during this project. Each of these failed samples was examined in detail to try and ascertain the reasons for failure, and the information was used in conjunction with other failure data compiled over the past 15 years to develop a defect database. The discussion below covers a representative sample of the failures examined.

The pipe shown in Figures 5.16 and 5.17 was installed in 1969 by an Australian authority, and failed in 2003. Initial examination of the outer surface of the pipe revealed a longitudinal fracture, approximately 15 mm in length. Due to discoloration of the outer surface, the failure is more clearly seen from the inner surface. Examination of the inner surface of the pipe revealed a longitudinal fracture, approximately 30 mm in length. The difference in visible fracture length indicates that the failure may have begun at or near the inner surface.

Figure 5.18 shows the exposed fracture surface of this pipe failure. The outer surface of the pipe is towards the top of the image and the inner surface towards the bottom. The observation plane is parallel to the pipe through-thickness direction. It can clearly be seen that the failure began at a cavity central to the pipe thickness. Semi-elliptical striations, or "beach marks", around the inherent cavity indicate that fracture initiated at this location, and grew radially towards the inner and outer surfaces of the pipe (Engel et al. 1978).

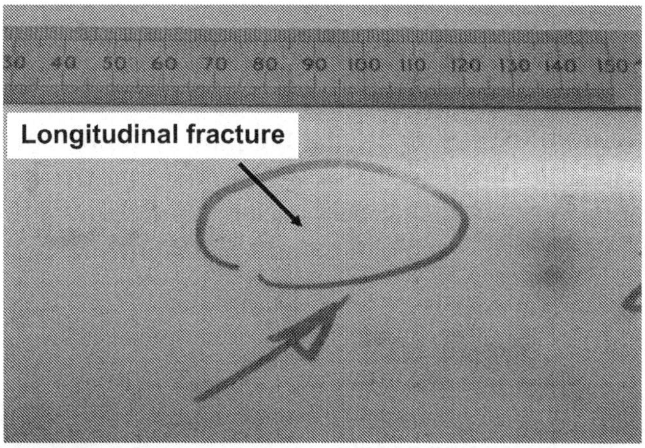

Figure 5.16 Outer surface of fractured pipe, Victoria, Australia: PVC-U DN 100, Class 12, Series I (installed 1969, failed 2003) (sample ID: 281699)

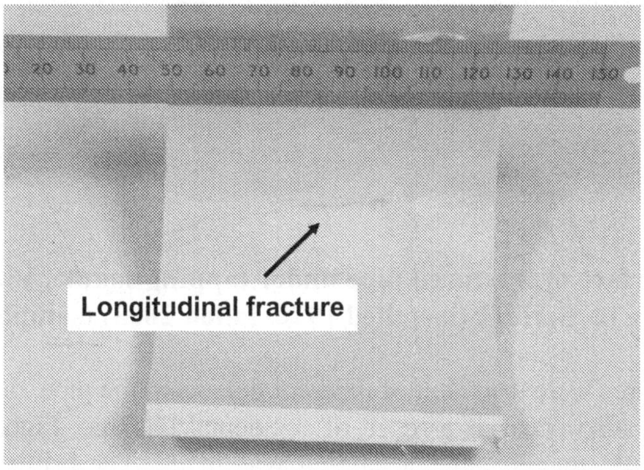

Figure 5.17 Inner surface of fractured pipe, Victoria, Australia: PVC-U DN 100, Class 12, Series I (installed 1969, failed 2003) (sample ID: 281669)

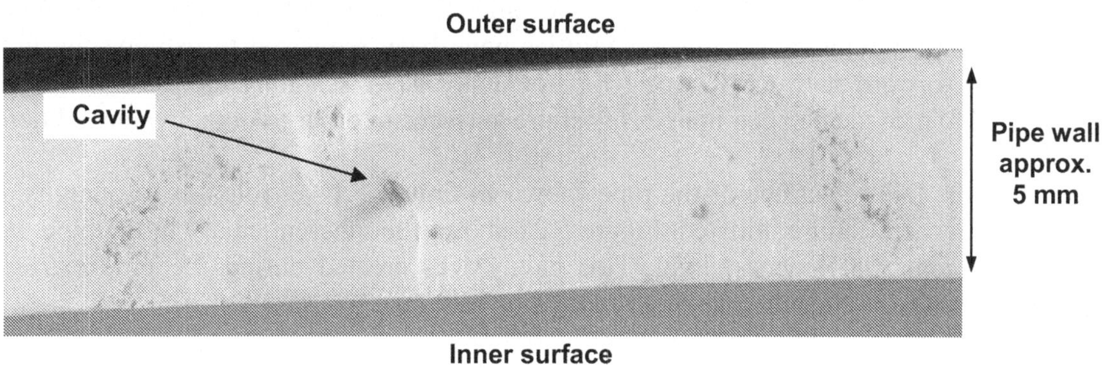

Figure 5.18 Fracture surface from PVC-U pipe sample (sample ID: 281669)

A second pipe (shown in Figure 5.19) was installed under a grass verge in 1970, and was reported as leaking in February 2003. As shown, the fracture was discovered underneath a tapping saddle, but was remote from the actual tapping hole (the tapping hole is situated approximately 90° away from the failure around the pipe circumference). The combined axial–circumferential alignment of the fracture indicates that the pipe was subjected to axial bending stresses in addition to hoop stress from internal pressure. A possible source of this bending stress was an overtightening of the saddle, which attempts to distort the circular cross-section of the pipe.

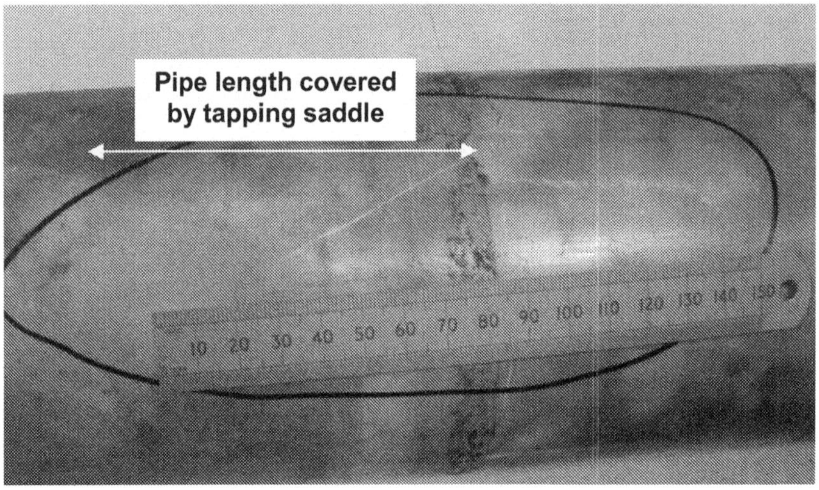

Figure 5.19 Outer surface of fractured pipe under tapping saddle, Victoria, Australia: PVC-U DN 100, Class 12, Series I (installed 1970, failed 2003) (sample ID: 273731)

These two failures were the result of inherent defects in the pipe, or due to poor installation, however, field failures also occur as a result of accidental damage. Figure 5.20 shows a PVC-U pipe that failed by brittle fracture after a tapping was made. The brittle fracture initiated at the tapping location shortly after tapping had been completed. Multiple occurrences of crack bifurcation and rejoining resulted in the removal of several sections of pipe wall.

In many cases, defects included at the time of manufacture were a source of initiation of fracture of PVC pipes. An indication of the size of defects that cause crack initiation is especially useful, since it provides input for the fracture mechanics failure model developed in Chapter 7. Whenever PVC pipe fractures occur, water utilities have been requested to exhume the failed pipe section and forward it to CSIRO or NRC for analysis. A scanning electron microscope (SEM) was used to provide detailed images of failure surfaces in order to obtain values of defect size.

SEM of the failure surface of the pipe shown in Figure 5.18 enabled us to quantify the size of the cavity where failure initiated. Figure 5.21 shows the inherent cavity in the pipe wall, approximately 1 mm (0.04 inches) long. This cavity was created during the pipe extrusion process, and may indicate the presence of air in the polymer melt.

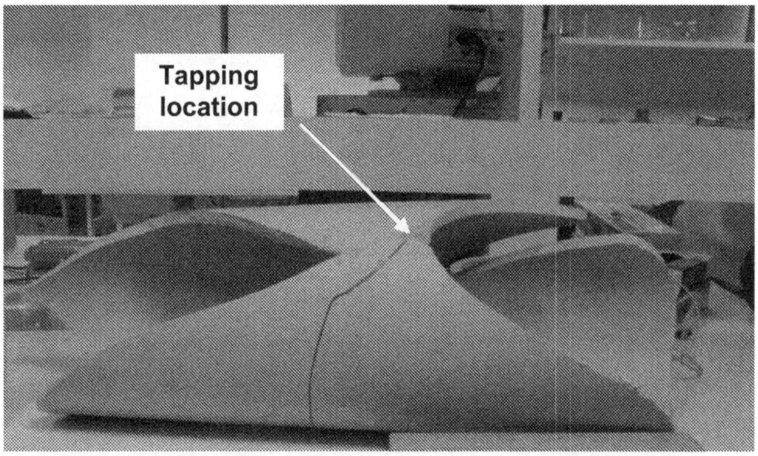

Figure 5.20 Tapping failure, New South Wales, Australia: PVC-U DN 200, Class 12, Series II (installation date unknown, failed 2000) (sample ID: unknown)

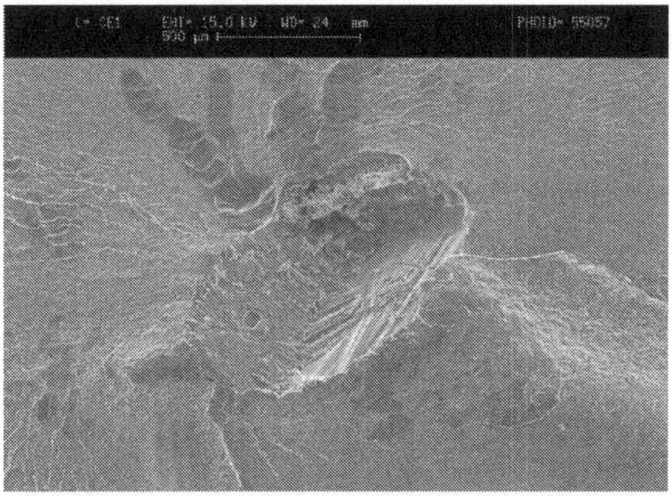

Figure 5.21 SEM image of fracture surface from PVC-U field failure (sample ID: 281699)

Figure 5.22 shows an SEM image of the fracture surface extracted from the field failure shown in Figure 5.19. Similar to the image in Figure 5.21, this cavity was probably due to an air bubble trapped in the polymer melt, which was not removed during the pipe extrusion process. However, in this case, the cavity is relatively small in size at 0.06 mm long.

A batch of failures in PVC pipes was recently reported in Tasmania, Australia, each occurring in PVC-U DN 300, class 12 pipes. The pipes were installed in 1993 and failures were recorded between 1997 and 2004. In each case, an axial crack was observed close to the socket on the pipe. Similar to previous failures, closer examination of this area revealed an inherent cavity, which may have caused crack initiation. In addition to cavities, failures in the Tasmanian PVC pipes were also initiated at particles included in the pipe wall, as shown in Figure 5.23.

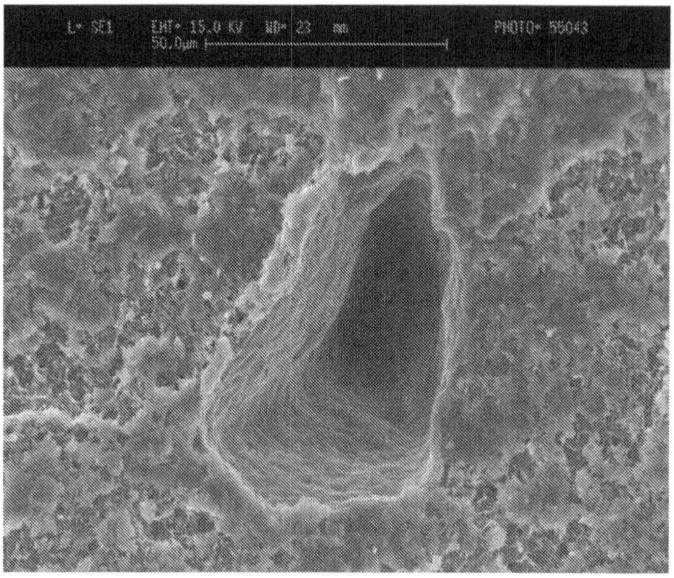

Figure 5.22 SEM image of fracture surface from PVC-U field failure (sample ID: 273731)

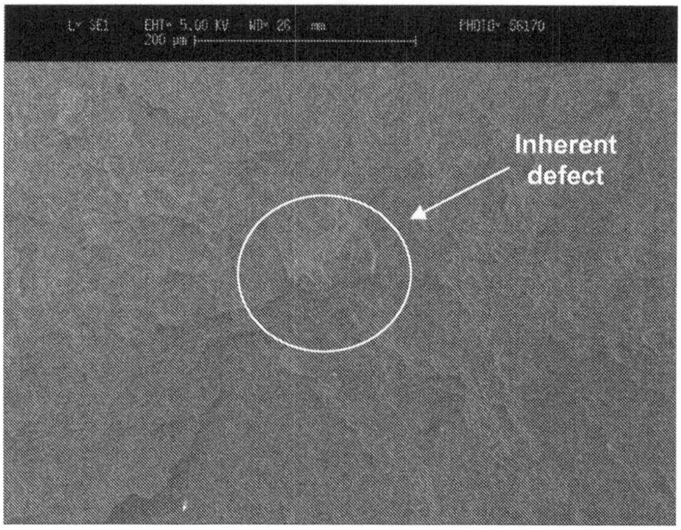

Figure 5.23 Particulate defect in fractured pipe, Tasmania, Australia: PVC-U DN 300, Class 12, Series II (installed 1993, failed between 1997 and 2004) (sample ID: TAS 5)

A foreign inclusion approximately 0.08 mm in diameter is clearly observed. An elemental analysis of this defect was conducted using energy dispersive X-ray spectroscopy (EDS). EDS is a standard procedure used to identify and quantify the elemental composition of samples. Areas as small as a few cubic micrometers can be analyzed independently of the rest of the sample. EDS of the particle and the surrounding bulk polymer was performed, with the resulting spectra shown in Figure 5.24.

The EDS spectrum reveals that the particle has an elemental composition significantly different from the bulk polymer, with the primary silicon and oxygen peaks indicating that the particle is likely to be a silicate of some type.

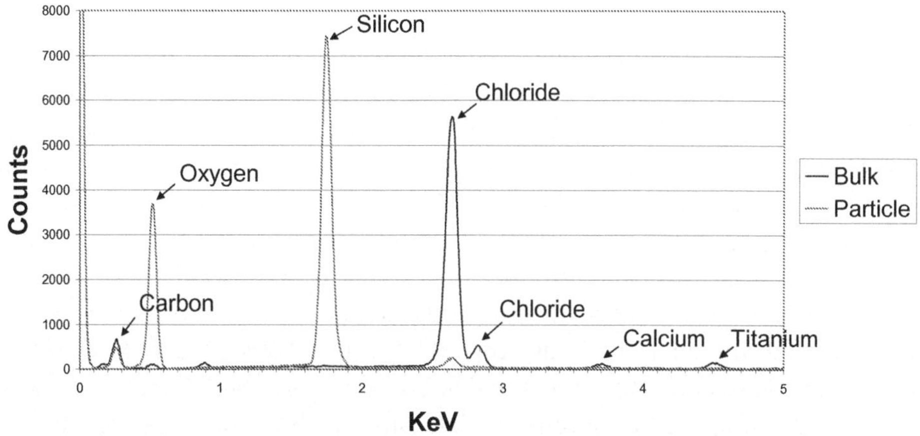

Figure 5.24 EDS spectra, Tasmania, Australia: PVC-U DN 300, Class 12, Series II (installed 1993, failed between 1997 and 2004) (sample ID: TAS 5)

CONCLUSIONS AND RECOMMENDATIONS

Responses from the performance survey indicate that PVC pipes are used to various extents within the water utilities. At the top, 41% of the City of Calgary Waterworks supply network length is made up of PVC pipes compared to only 2.4% of the South Australia Water supply network length at the bottom. It should be noted, however, that despite the small percentage, South Australia still has a considerable length of installed PVC pipe.

Of those water utilities that supplied failure data, a range of annual failure rates (number/100km/year) was estimated. At the top of this range, South East Water exhibited an estimated annual failure rate of 6.8 failures/100km/year. At the bottom of the range, Epcor exhibited an estimated annual failure rate of 0.04 failures/100km/year. However, it should be reiterated that these figures are merely estimates based on the total number of failures recorded over a time period, and the current length of PVC pipe. In practice, however, both the length of PVC pipe and the number of failures experienced will vary from year to year. An effective method of recording and organizing pipe asset and failure data is therefore recommended to capture this variation. Until data can be represented in this manner, care must be taken in interpreting and comparing failure rates. It should also be noted that in most cases the number of recorded failures in PVC pipes was low compared to other materials. For example, EBMUD recorded only 21 failures in PVC pipes over the period 1996–2002. Furthermore, in 2002, the average age of PVC pipes in the EBMUD system was 9 years, with a maximum age of approximately 20 years. Consequently, expecting the breakdown in failure types, as reported by this survey, to be representative of PVC pipes as they age may not be realistic. However, it is still informative to compare current trends in failure types with other water utilities.

Examining the failures sustained in individual water utilities illustrated the dominant failure modes in each case. Since it is likely that the external factors governing pipe failure (i.e., ground movement) acted in different magnitudes for each water authority, different dominant failure modes were observed in each case. In general, joint leakage appeared to be a problem for EBMUD. For the City of Ottawa, the fraction of longitudinal and circumferential fractures was relatively high. This was attributed to relatively low ground temperatures in Canada, which can induce direct axial tensile stresses (by frictionally restrained axial pipe contraction), and contribute to circumferential stresses (by frost loading and diametrical deflection of pipes). Further information is required to identify the dominant failure modes for Hunter Water. However, the relatively low percentage of circumferential fractures was attributed to higher ground temperatures, which would not induce direct axial tensile stress. Pipe flexibility and the extent of pipe lateral movement accommodated in rubber-ring joints would also reduce axial bending stresses, which can result in circumferential fractures. For Yarra Valley Water, fitting failures were dominant. Further information is required to determine whether this corresponds to fracture of the fitting itself or leakage from the pipe/fitting interface. As with Yarra Valley Water, a large proportion of failures in South East Water were attributed to tapping fittings. The slightly lower proportion of circumferential fractures for South East Water (9.9% compared to 13% for Yarra Valley Water) was attributed to a relatively small area of expansive clay soils in the South East Water network.

For comparison, failure data recorded in 1992 and 1993 for PVC pipes in Canadian water utilities was obtained. This data indicated a significantly lower average failure rate for Canadian utilities, ranging from 0.9 to 3.0 failures/100km/year. A comparison of failure modes also indicated that Canadian utilities exhibited a greater fraction of longitudinal fracture failures in

PVC pipes than Australian utilities. While the action of frost loads on Canadian pipes was proposed as a possible reason for this difference, more failure data are required to confirm this.

For comparison with Australian, and USA/Canadian PVC pipe failures, a national failure database was supplied by UK Water Industry Research (UKWIR). The average failure rate in UK PVC-U pipes is comparable (if slightly higher) to that reported by Australian water utilities, but significantly higher than the range reported by USA/Canadian utilities. 27% of recorded failures in UK PVC-U pipes are classified as "unknown", followed by 26% classified as "pipe". It is unclear which particular PVC pipe failure mode this description refers to. 20% of recorded failures are described as "joint" which probably corresponds to leaking joints, and 17% of failures are classified as "longitudinal" corresponding to longitudinal fracture of a pipe. While 3% of failures are classified as "Pin Hole", this probably corresponds to very short longitudinal fractures, created by slow crack growth through the pipe wall (a similar failure mode "perforation" was reported by Yarra Valley Water (Australia)). Only 3% of recorded failures are classified as "circumferential" corresponding to circumferential fracture. This may reflect the extent of lateral pipe movement afforded by rubber ring joints in PVC pipes. Both USA/Canadian and Australian water utilities exhibit a similar fraction of longitudinal fracture failures (30 and 36.3% respectively). A lower percentage (20%) was reported by UK water utilities, which may be attributed to the significant fraction of failures recorded as "unknown" and "pipe". It is possible that a number of these failures occurred by longitudinal fracture.

An analysis of the UKWIR database allowed the variation in average failure rate with pipe age to be extracted. For both fracture failures and joints/fittings failures, a relatively constant average failure rate is observed until a sharp increase at a pipe age of around 42 years old. This sharp increase in fracture failure rate for older pipes is accredited to PVC pipe production techniques and poor installation techniques, both of which improved when revised practices were introduced in the 1970's. Fracture failures in PVC pipes installed before 1965 to 1970 were attributed to poor control of pipe processing which resulted in low fracture toughness levels. For pipes installed after 1971, processing control and fracture toughness levels improved, which would reduce the number of fracture failures observed in the field. Improved installation practices introduced in 1973 would explain the improvement in joint/fitting failures observed.

A set of samples from fracture failures in Australian PVC pipes has been examined. In most cases, crack initiation has occurred at inherent defects in the pipe wall. These may be attributed to cavities from unfilled air bubbles during the pipe manufacturing process, or foreign particles (dust or grit) that were included during pipe extrusion. Crack initiation occurred at, or near to, the inside surface of the pipe and then propagated through the wall thickness, finally penetrating the pipe outer surface. Longitudinal fractures which were relatively short in length were observed in most cases. This may indicate that crack growth occurred under normal operating conditions, in the absence of extreme loading events, such as pressure surges, for example. SEM examination of fracture surfaces also indicated that crack growth was usually accompanied by local ductility, and the resulting fracture was not purely brittle.

The variability in failure modes from authority to authority indicates that installation practices may be a much larger factor in determining the types of failure modes than pipe manufacture. Consequently, collection of failure data and detailed analysis of the data using consistent formats could lead to early detection of factors affecting premature failure of PVC pipes. However, in comparison, examination of failure surfaces from the limited number of collected pipes indicated that, of the failures examined, manufacturing defects played a role in premature failure.

CHAPTER 6
FRACTURE PROPERTY ASSESSMENT OF PVC PIPES

The field performance survey of PVC pipes indicates that a range of failure types occur in service. For example, Yarra Valley Water recorded 384 failures in PVC-U pipes in the period 1995–2003. Of these, 29% were attributed to fitting failures; 19% were attributed to longitudinal fractures; 18% of failures were described as other; 13% were classified as circumferential fractures, and 10% of failures were attributed to blown sections. The remaining failure types were classified as damaged (4%), perforations (3%), unknown (3%), and joint leaks (1%). From this total, longitudinal fractures, circumferential fractures, blown sections, and perforations can all be investigated further using fracture mechanics modeling, with other types of failures being addressed principally by improving installation conditions.

As shown in Chapter 5, fracture failures in service often occur from inherent defects in the pipe wall. Under internal pressure, deflection, and other service loads, cracks can initiate from these defects and grow until they become unstable and "jump" through the pipe wall.

To develop a model to predict when these failures can occur in service, several important fracture properties of the pipe material (in this case PVC) are required. This chapter presents test methods used to determine these properties, and results from tests on different PVC pipe types. Comparing these properties gives an indication of those pipes that are more likely to fail by fracture in the field. These properties are also essential input to a failure model for PVC-U pipes, which is described in detail in Chapter 7.

FRACTURE TEST METHODS USED

C-Ring Test for Plane-Strain Fracture Toughness

The C-ring test method is widely used by the water industry to determine the plane-strain fracture toughness of PVC pipes. The test currently forms a draft ISO standard (ISO/DIS 11673.2), and an interim Australian/New Zealand Standard (AS/NZS 1462.19(Int)) for fracture toughness assessment.

As shown in Figure 6.1, a split ring section is machined from a length of pipe, and a sharp notch is introduced from the inner pipe surface to partway through the ring (pipe wall) thickness. A static load is then attached to the lower half of the ring, applying a flexural stress, which attempts to open the notch. Under these conditions, brittle fracture can be induced in a similar manner to the bending action of a point load at the outer surface of a buried pipe. Figure 6.2 shows a set of C-ring tests in progress. Crack initiation and the onset of failure can be seen in Figure 6.3.

As outlined in AS/NZS 1462.19(Int), the SIF K_I associated with a notched C-ring is written in terms of the applied bending moment M:

$$K_I = \frac{6YM\sqrt{\pi a}}{Wb^2} \tag{6.1}$$

where Y is a geometric correction factor, W is the width of the C-ring, b is the ring (pipe wall) thickness, and a is the initial notch depth. LEFM theory proposes that unstable fracture will occur when a critical SIF K_{IC} is attained, given by (the subscript "c" denotes critical values):

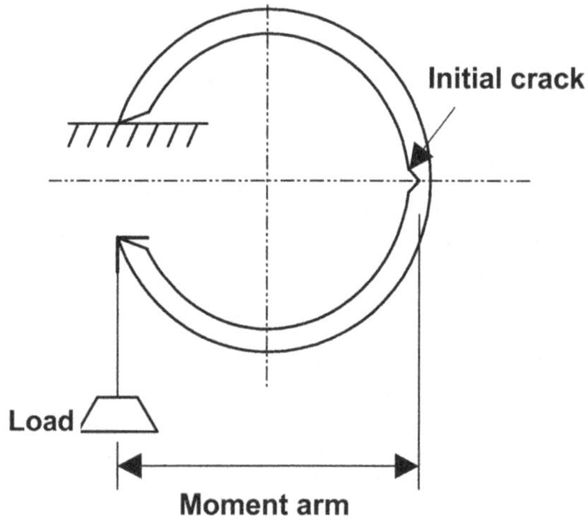

Figure 6.1 C-ring fracture toughness test

Figure 6.2 C-ring testing in progress

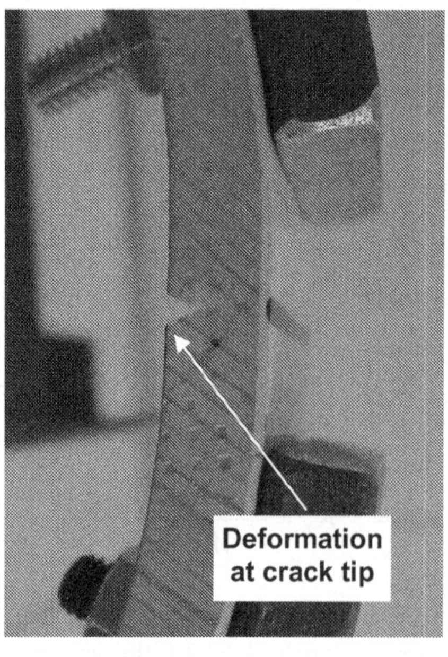

Figure 6.3 Crack initiation and notch plasticity in the C-ring test

$$K_{IC} = \frac{6YM_C\sqrt{\pi a_c}}{Wb^2} \tag{6.2}$$

Compared to other specimen geometries, the wall thickness of PVC pipe is below the ASTM size criteria for plane-strain conditions (ASTM D5045). LEFM theory must therefore be modified, as yielding extends from the notch tip, resulting in a non-linear response to the applied load. To solve this problem, fracture in the post-yield regime is represented using the BCS yield theory (Bilby, Cottrell, and Swinden 1963). The result is a correction to the critical applied SIF that accounts for this plastic deformation (AS/NZS 1462.19(Int)).

Since the C-ring test is a "*K*-increasing" test (Ewalds and Wanhill 1989), all three stages of the fracture failure process are observed. This is shown schematically in Figure 6.4.

An incubation period is observed at the initial applied SIF, K_{Ii}, followed by crack initiation. A period of slow crack growth then follows, and the SIF increases with increasing crack length. At some point, $K = K_{IC}$ (the true plane-strain fracture toughness) the criterion for fast brittle fracture is satisfied. While the true fracture toughness K_{IC} can be extracted from this test, the results primarily indicate the relationship between an *initial* applied SIF, and the time to failure. Visual examination of a fracture surface from the C-ring tests indicates the three-stage failure process (Figure 6.5).

As shown, the region ahead of the notch shows stress whitening, indicating slow crack growth after crack initiation. A relatively smooth, featureless region is then observed, indicating the transition to fast brittle fracture.

SENB Test for Slow Crack Growth Resistance

A simple method for determining the incubation period and slow crack growth resistance is the single edge notch bend (SENB) test (Figure 6.6).

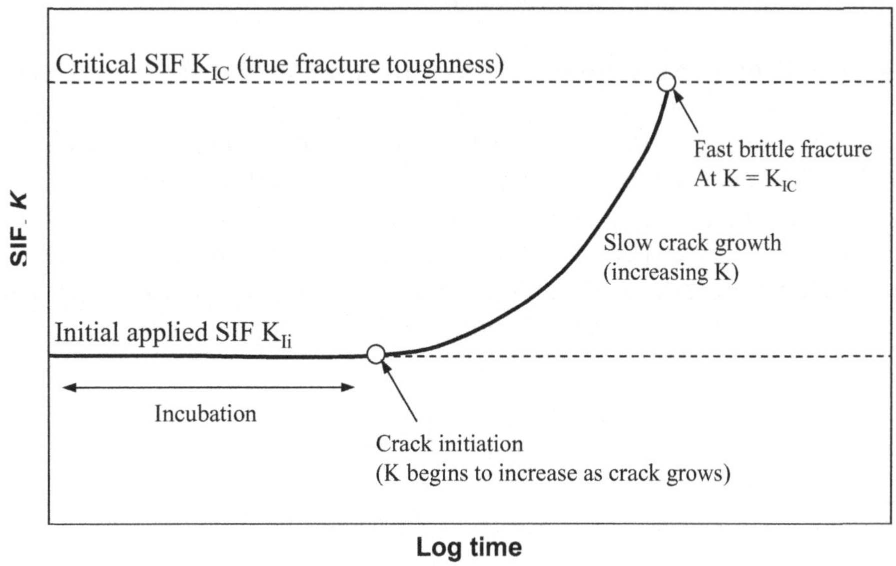

Figure 6.4 Increasing SIF in C-ring test

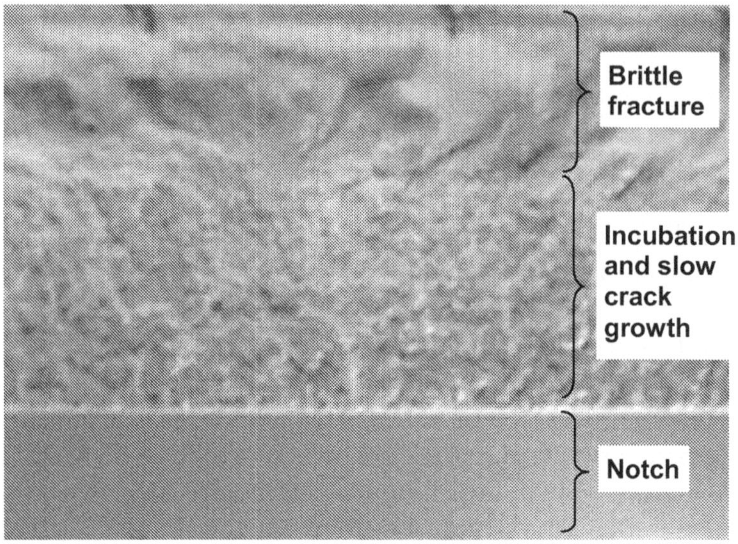

Figure 6.5 Fracture surface from C-ring test on PVC-U pipe

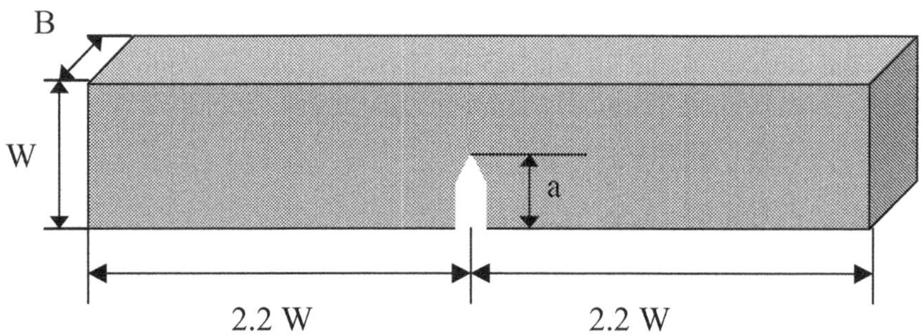

Where: W=2B and 0.45<a/w< 0.55

Figure 6.6 Schematic of SENB sample for crack incubation and slow crack growth testing

As shown in Figure 6.6, the SENB specimen is a rectangular bar, with a v-shaped notch at the underside. The specimen is supported at its ends (point supports), and a static load is applied to the surface directly opposite the notch. The loading mode is three-point bending; *a* is the depth of the notch; *B* is the specimen thickness; and *W* is the specimen width. To ensure that plane-strain conditions prevail, *a*, *B*, and *W* must be chosen, as shown in Figure 6.6. As prescribed in ASTM D5045, the applied SIF for an SENB sample is given by:

$$K_I = \left(\frac{PS}{BW^{3/2}} \right) f\left(\frac{a}{W} \right) \tag{6.3}$$

where *P* is the applied load, and *S* is the specimen span. The function *f*(*a*/*W*) is a geometric correction factor.

As outlined previously, other researchers have proposed that incubation ends when a critical crack opening displacement (COD) is attained, which is independent of specimen or component geometry (Williams 1978, Marshall 1982). For a line plastic zone ahead of a crack, this critical COD, δ^*, can be written as:

$$\delta^* = \frac{K_I{}^2 t_i{}^{(p+q)}}{E_0 \sigma_0} \tag{6.4}$$

where K_I is the applied SIF, and visco-elastic behavior is represented by the time-dependant Young's modulus $E = E_0 t^{-p}$, and yield stress $\sigma = \sigma_0 t^{-p}$. t_i is the incubation time between application of an initial SIF and crack initiation.

Monitoring the time-dependant compliance of an SENB sample allows t_i, and hence δ^*, to be determined (Bradley, Slay and Self 1998). Substituting the incubation time observed from a particular SENB test back into Equation 6.4 gives the critical COD required for crack initiation. Assuming that this δ^* value is a material constant (Williams 1978, Marshall 1982), this then allows incubation times for all specimen and component geometries to be determined by rearranging Equation 6.4:

$$t_i = \left(\frac{E_0 \sigma_0 \delta^*}{K_I{}^2} \right)^{\frac{1}{p+q}} \tag{6.5}$$

This equation is applied to determine incubation times from SENB tests later in this chapter. After incubation ends, a period of slow steady crack growth occurs, followed by a period of unsteady crack growth when the crack becomes unstable as K_I approaches K_{IC}. Referring back to Equation 3.4, K_I can be related empirically to crack growth rate in the form:

$$\frac{da}{dt} = D K_I{}^m \tag{6.6}$$

where D and m are constants for a particular environment. Following previous researchers (Marshall 1982), the crack growth rate in a loaded specimen can be obtained from the time-dependant changes in specimen compliance. By separating material compliance from total compliance, compliance changes due to crack growth alone can be obtained (Bradley, Slay and Self 1998). In turn, these can be related to changes in crack length via the fracture mechanics compliance relationship for an SENB specimen (Atkins and Mai 1988). Changes to K_I with crack length can be obtained from Equation 6.3. The SENB test can therefore be used to determine the incubation period, critical COD, and the slow crack growth rate parameters for a particular material. These calculations are presented later in this chapter.

EWF Test for Plane-Stress Fracture Resistance

In contrast to the previous test methods, the EWF test is used to quantify resistance to ductile fracture. There is increasing use of the method to determine the toughness of thin plastic films, and the method is attractive in that it deals with the sort of gross ductility that can occur in the plane-stress fracture state. Physically this means that the observed failure mode in an EWF test is one of ductile yielding and tearing fracture, rather than brittle fracture. As will be seen later, the test is suited to the analysis of PVC-M, which exhibits increased ductility and post-yield fracture.

The EWF approach is a means of partitioning the energy associated with fracture into two parts. One part is specific to the fracture of the material and, as such, is a material property. The second part is related to gross plastic deformation and depends on the geometry of the fractured specimen or component. The basis of this partitioning was originally suggested by Broberg (1968). He proposed that the non-elastic region at the tip of a crack should be subdivided into an inner region where the fracture process takes place, and an outer region surrounding it where gross ductility occurs (Figure 6.7). The crack tip-specific work is defined as the essential work of fracture (W_e), and the work done in the outer zone as the non-essential work of fracture (W_p).

The total fracture work (W_f) can by expressed as:

$$W_f = W_e + W_p \tag{6.7}$$

W_e is proportional to the length of the ligament, and W_p is proportional to the square of the ligament (Broberg 1968). Equation 6.7 can therefore be expressed as:

$$W_f = w_e lt + \beta w_p l^2 t \tag{6.8}$$

where l is the sample ligament length, t is the sample thickness, β is a geometry-dependant factor related to the plastic zone, w_e is the specific essential work of fracture, and w_p is the specific non-essential work of fracture.

Dividing Equation 6.8 by the cross-sectional area of the ligament (lt) gives the specific total work of fracture w_f as:

$$w_f = w_e + \beta w_p l \tag{6.9}$$

Therefore, if Equation 6.9 is valid, plotting the variation of w_f with l should give a straight line, with an intercept w_e and a slope equal to βw_p.

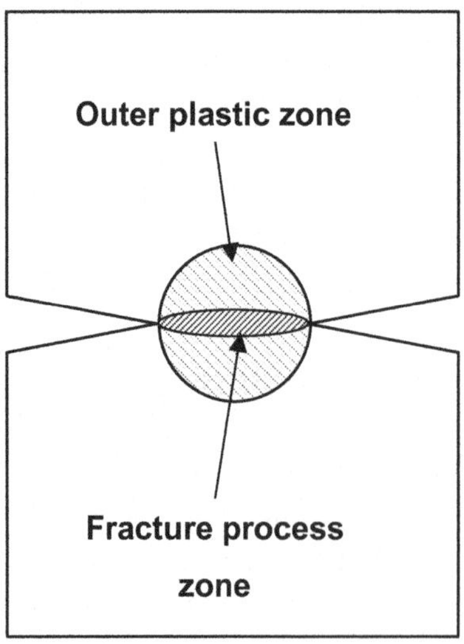

Figure 6.7 EWF fracture test specimen

The EWF sample must be under plane-stress conditions and the ligament must be fully yielding before crack propagation. The transition between plane-strain and plane-stress states can be difficult to define (Clutton 2001). However, according to the literature (Clutton 2001, Levita et al. 1996, Lau and Truss 2002), the lower limit for plane stress is achieved when the sample dimensions are such that the ratio of ligament length to specimen thickness (l/t) is between 3 and 5. For ratios of l/t less than 3, deformation would occur under a mixed plane-stress/plane-strain mode, and deriving EWF becomes difficult. The upper limit or maximum ligament size is defined as the minimum between $W/3$, or two times the plastic zone length ($2r_p$) (Clutton 2001). However, such a definition is considered arbitrary and, for practical purposes, upper limits have been defined as 15 mm, or any other ligament values, as long as the ligament is fully yielded upon failure (Clutton 2001).

An EWF test in progress and a failed specimen are shown in Figures 6.8 and 6.9 respectively. As shown in Figure 6.9, a fully yielded zone encompasses the ligament at failure. Note also the local thinning of the ligament thickness during the test. This indicates that necking and drawing of material accompanies fracture.

Rotational Bending Test for Fatigue Resistance

As outlined by Whittle (2002), cyclic fatigue testing was conducted using cylindrical rod specimens, which were notched to a predetermined depth around their circumference. These rods were mounted in a cantilever arrangement, and were rotated in a motorized chuck. The numbers of cycles to failure were recorded. Figure 6.10 shows a notched rod held in the chuck jaws.

The rotational bending apparatus is shown in Figure 6.11.

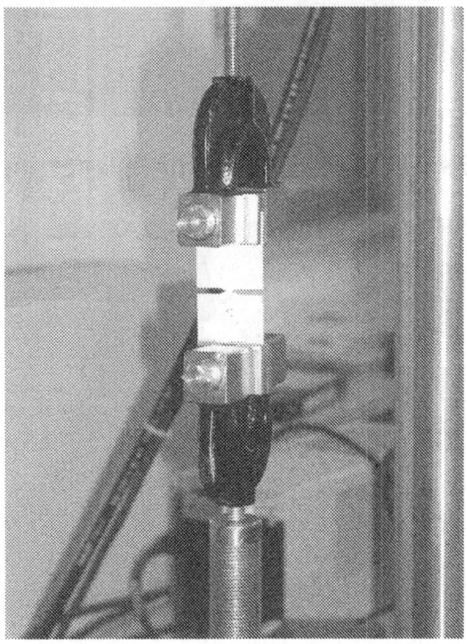

Figure 6.8 EWF test in progress

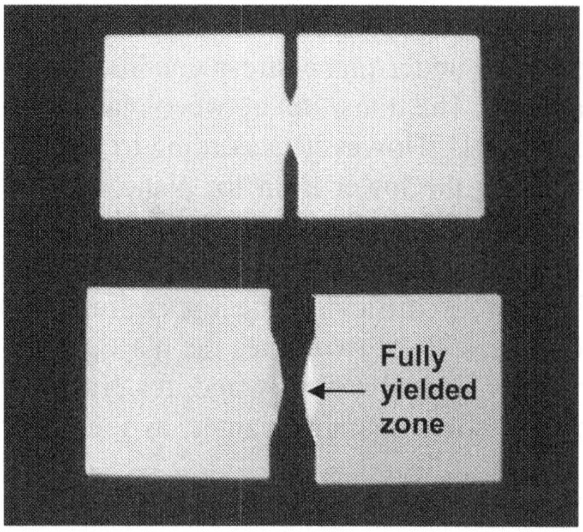

Figure 6.9 EWF failed specimen

Figure 6.10 Notched rod specimen for cyclic fatigue testing

Figure 6.11 Rotational bending test arrangement for cyclic fatigue testing

As shown, the specimen cantilever arrangement is shown on the left of the picture. The suspended mass on the right of the picture exerts an upwards force at the end of the specimen via the frame. The end of the rod is supported on bearings and the motor is used to rotate the chuck, which applies a cyclic load to each point around the circumference at the end of the cylindrical rod specimen. In turn, this applies a cyclic bending moment to the notched region of the specimen, which can be converted to an SIF range Δ*K*.

According to Whittle (2002), the applied SIF range for the rotational bending specimen is given by:

$$K = \sigma\sqrt{\pi a}F\left(\frac{a}{b}\right) \tag{6.10}$$

where *a* is the radius of the ligament at the notch, *b* is the radius of the rod, and *F*(*a*/*b*) is a geometric correction factor depending on the notched ligament and rod radii. For the bending arrangement shown in Figure 6.11, the applied stress at the notch, σ, is given by:

$$\sigma = \frac{4ml}{\pi a^3} \tag{6.11}$$

where *m* is the applied load, and *l* is the distance from the load point to the notch. Under cyclic conditions, the applied SIF range becomes:

$$\Delta K = \Delta\sigma\sqrt{\pi a}F\left(\frac{a}{b}\right) \tag{6.12}$$

C-RING TEST RESULTS – PLANE-STRAIN FRACTURE TOUGHNESS

The details of C-ring fracture toughness test specimens are given in Table 6.1. The variation in K_{Ii} with time is plotted in Figure 6.12 for the range of PVC-U samples. In each case, a power-law curve of best fit has also been applied (corresponding to the "Power" notation in the legend).

Table 6.1
C-ring fracture toughness testing: sample details

Sample	Nominal diameter	Pressure class MPA (psi)	OD mm (inches)	Wall mm (inches)
PVC-U Ca/Zn (AUS)	DN 100 mm	1.6 (231.9)	122.2 (4.81)	8.8 (0.35)
PVC-U Lead (AUS)	DN 100 mm	1.8 (260.9)	122.2 (4.81)	9.8 (0.39)
PVC-U Tin (AUS)	DN 100 mm	1.2 (173.9)	114.3 (4.50)	6.3 (0.25)
PVC-U Tin (USA)	DN 4 inches	0.95 (150)	122.0 (4.80)	7.1 (0.28)
PVC-M Ca/Zn (AUS)	DN 100 mm	1.6 (231.9)	121.9 (4.80)	6.1 (0.24)
PVC-M Lead (AUS)	DN 100 mm	1.8 (260.9)	114.3 (4.50)	6.3 (0.25)
PVC-M Tin (AUS)	DN 100 mm	1.8 (260.9)	114.3 (4.50)	6.3 (0.25)

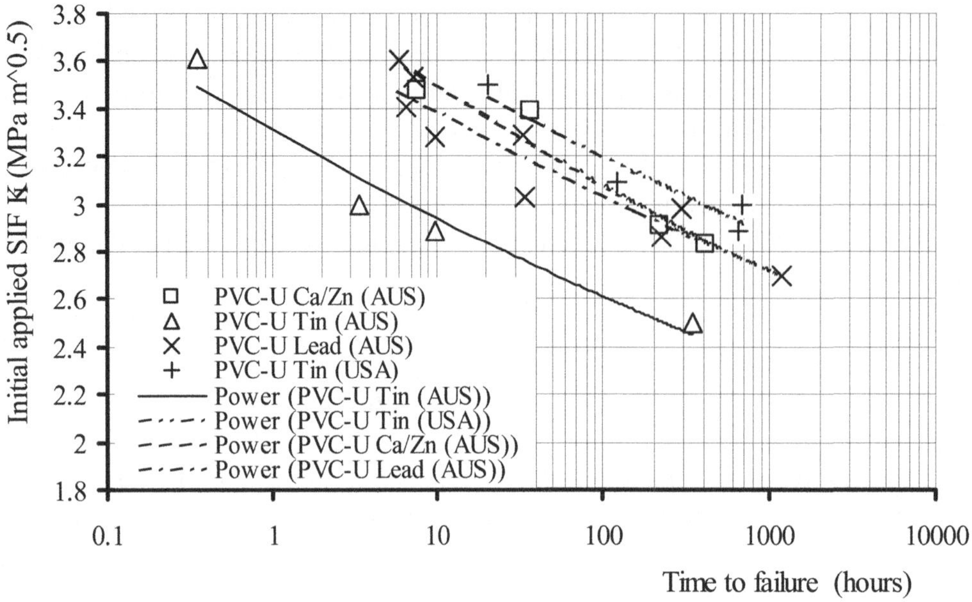

Figure 6.12 K_{Ii} versus failure time for different PVC-U pipe samples: C-ring tests

As shown in Figure 6.12, the data seems to follow a similar trend in each case: as initial applied SIF decreases, failure time increases approximately linearly on a logarithmic scale. Apart from PVC-U Ca/Zn, K_{Ii} values also appear to decrease at similar rates, and the difference between stabilizer types is reflected in a vertical shift of the K_{Ii}/time plot. Across the range of recorded failure times, the results in Figure 6.12 indicate that Sn-stabilized USA-manufactured PVC-U pipe is superior. This is followed by Ca/Zn- and Pb-stabilized PVC-U (each with similar rates of reduction in K_{Ii}), and Sn-stabilized PVC-U pipe manufactured in Australia.

The linear time dependence on the logarithmic scale suggests that an empirical power-law time dependence in the form:

$$K_{Ii}(t) = K_o t^{-Q} \tag{6.13}$$

can be applied to the data (K_{Ii} is in MPa.m$^{0.5}$, and t is in hours). Table 6.2 lists the K_0 and q empirical parameters obtained from fitting the data for each PVC-U sample.

Table 6.2
Power-law empirical parameters describing K_{Ii} versus failure time data for PVC-U samples: C-ring tests

Sample	K_0	Q
PVC-U Ca/Zn (AUS)	3.99	–0.056
PVC-U Lead (AUS)	3.78	–0.048
PVC-U Tin (AUS)	3.31	–0.051
PVC-U Tin (USA)	4.00	–0.048

The Australian/New Zealand performance requirements for PVC-U pressure pipes (AS/NZS 1477) stipulate that the initial applied SIF to produce failure at 15 minutes should exceed a minimum value that depends on pipe size and pressure class. Since periods of crack incubation and crack growth are relatively short, the K_{Ii} value that produces failure within 15 minutes is referred to as the plane-strain fracture toughness, K_{IC}. Table 6.3 compares values of K_{IC} extrapolated from Equation 6.13 at 15 minutes with those stipulated in AS/NZS 1477.

As shown, all samples pass the fracture toughness performance requirements stipulated in AS/NZS 1477. It is interesting to note that in terms of short-term K_{IC} (at 15 minutes), PVC-U Ca/Zn (AUS) performs slightly better than PVC-U Tin (USA). However, referring back to Figure 6.12, this superior performance is not observed for longer failure times. Following the previous discussion, the superior long-term performance of PVC-U Tin (USA) should be reflected in its crack incubation and slow crack growth properties. This hypothesis is explored in the next section.

For comparison, the variation in K_{Ii} with time is plotted for a range of PVC-M samples in Figure 6.13.

Table 6.3
Comparison of K_{IC} at 15 minutes for PVC-U samples with the performance requirements in AS/NZS 1477

Sample	Size	Class	K_{IC} at 15 mins ($MPa.m^{0.5}$) (Equation 6.13)	K_{IC} at 15 mins ($MPa.m^{0.5}$) (AS/NZS 1477)	Pass
PVC-U Ca/Zn (AUS)	DN 100 mm	1.6 MPa (231.9 psi)	4.31	3.75	Y
PVC-U Lead (AUS)	DN 100 mm	1.8 MPa (260.9 psi)	4.04	3.75	Y
PVC-U Tin (AUS)	DN 100 mm	1.2 MPa (173.9 psi)	3.55	3.30	Y
PVC-U Tin (USA)	DN 4 inches	150 psi (0.95 MPa)	4.25	n/a	Y

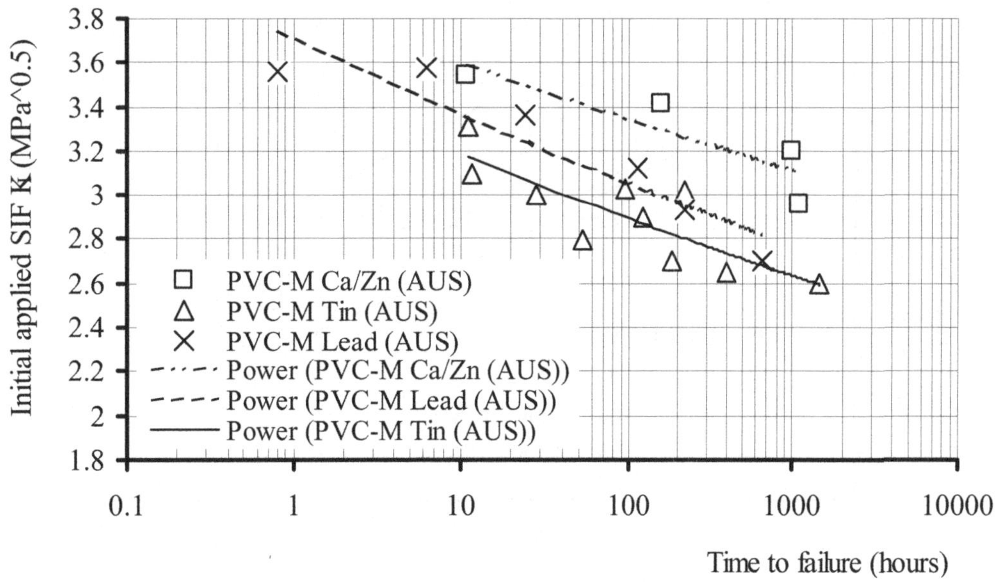

Figure 6.13 K_{Ii} versus failure time for different PVC-M pipe samples

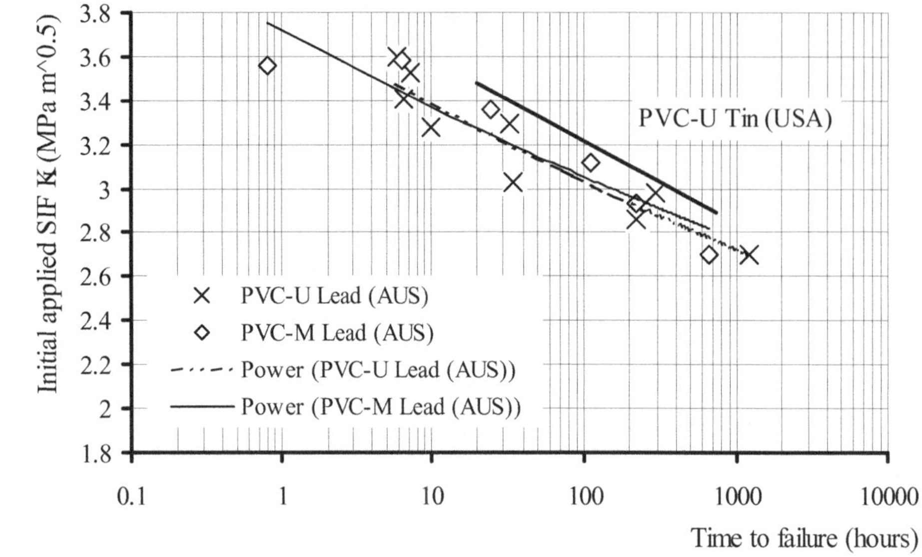

Figure 6.14 Comparison of initial applied SIF K_{Ii} versus time to failure for PVC-U and PVC-M lead-stabilized pipe samples

Similarly to PVC-U, linear time dependence is observed on the logarithmic scale for PVC-M, and the difference between stabilizer types is reflected in a vertical shift of the data. Across the range of recorded failure times, the results in Figure 6.13 indicate that Ca/Zn-stabilized Australian-manufactured PVC-M pipe is superior. This is followed by Sn- and Pb-stabilized pipes manufactured in Australia.

Plots of K_{Ii} versus time to failure for PVC-U and PVC-M samples are compared in Figures 6.14–6.16. For comparison, the data for PVC-U Tin (USA) is also included in each plot.

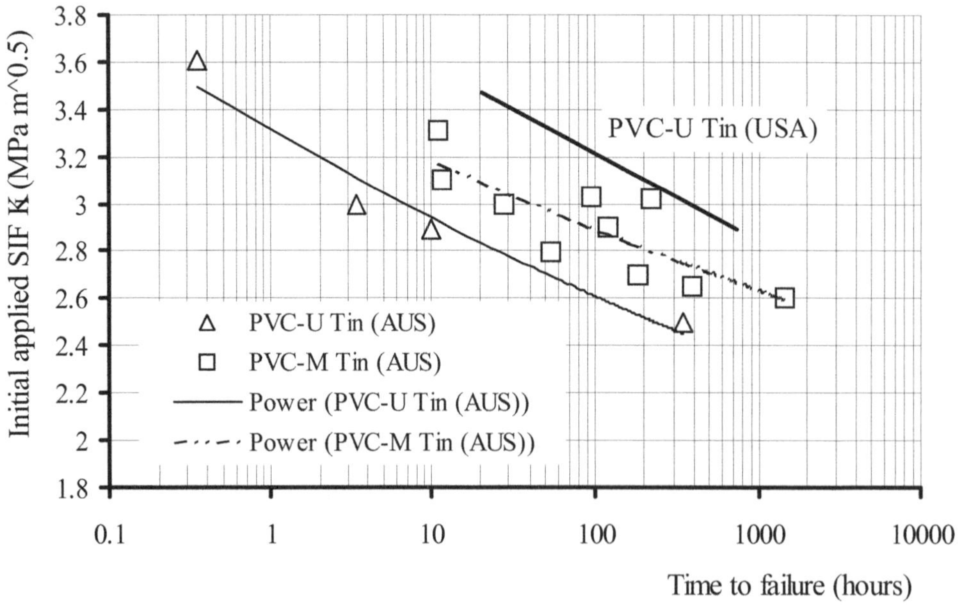

Figure 6.15 Comparison of K_{Ii} versus time to failure for PVC-U and PVC-M Sn-stabilized pipe samples

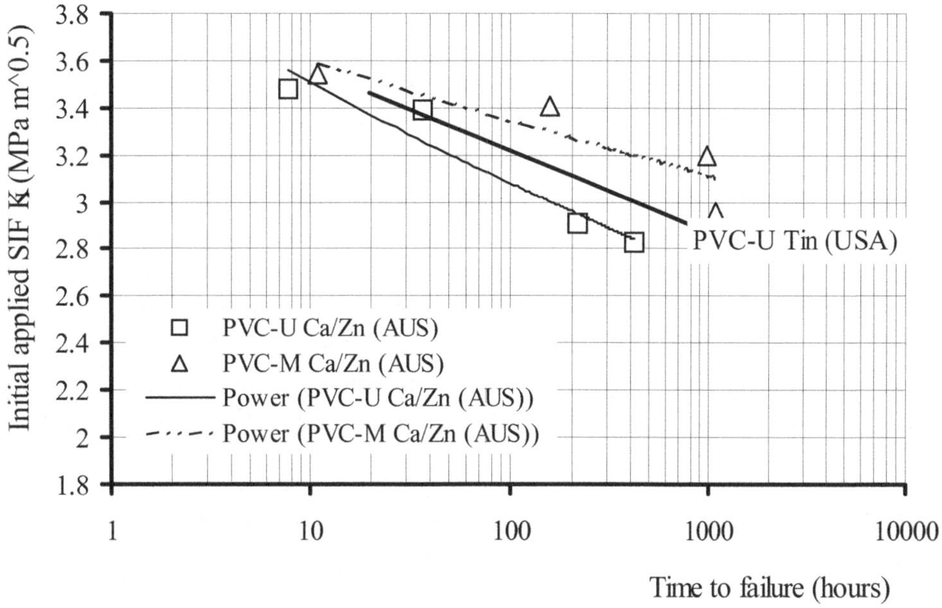

Figure 6.16 Comparison of K_{Ii} versus time to failure for PVC-U and PVC-M Ca/Zn-stabilized pipe samples

As shown, the general trend is that PVC-M exhibits a longer failure time for given initial applied SIFs. However, only PVC-M Ca/Zn (AUS) exhibits a superior performance to PVC-U Tin (USA).

While initial SIFs can be plotted for PVC-M samples, examination of fracture surfaces indicates that the application of LEFM theory is debatable (Figure 6.17).

For PVC-U (left image), the whitened region ahead of the notch corresponds to slow crack growth through a crazed zone. Although ductile deformation has occurred, its extent is limited. A clear transition in failure mode occurs at a critical crack length (approximately 4 mm in this case). The subsequent non-whitened surface corresponds to crack acceleration and fast brittle fracture through the remaining specimen ligament.

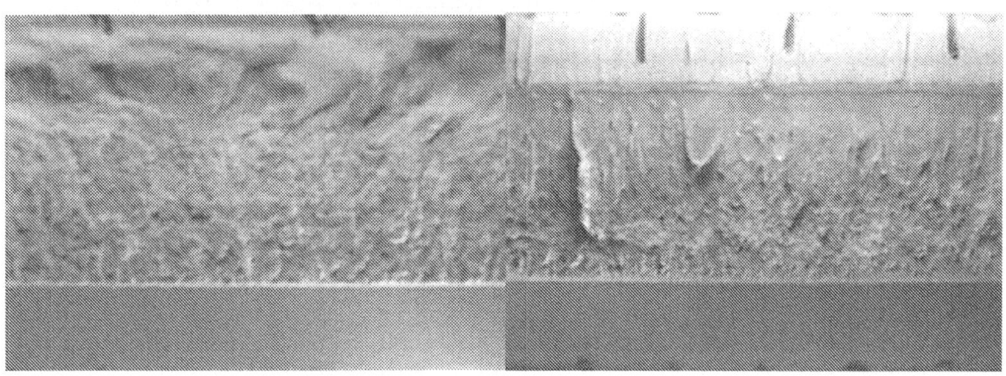

Figure 6.17 Comparison of fracture surfaces from C-ring tests on PVC-U Ca/Zn (AUS) (left) and PVC-M Ca/Zn (AUS) (right)

The PVC-M specimen (right image) has failed by slow crack growth in the post-yield regime (indicated by the enhanced whitening on the surface). Ductile tearing is evidenced by the v-shaped "chevrons" pointing in the opposite direction to that of tearing propagation (Engel et al. 1978). A plastic hinge at approximately 4 mm from the notch tip indicates eventual plastic collapse of the specimen ligament. The absence of brittle fracture in the PVC-M specimen is consistent with the addition of a ductility-promoting modifier, such as CPE. Since the failure mechanism in PVC-M samples in the C-ring test is not brittle fracture, the variation of initial SIF with time to failure cannot be used as a failure criterion. It is interesting to note that, although PVC-U Tin (USA) exhibits longer failure times, its fracture surface (Figure 6.18) is similar to that of PVC-U Ca/Zn (AUS) (Figure 6.17). However, it does exhibits greater stress whitening in the slow crack growth region. While this indicates an increase in local plastic deformation, the failure modes are comparable. Again, a sharp transition is observed at a crack length of approximately 4 mm, where fast brittle fracture propagates through the remaining sample thickness.

SENB TEST RESULTS – CRACK INCUBATION AND SLOW CRACK GROWTH RESISTANCE

As described previously, the SENB test can be used to determine crack incubation and slow crack growth resistance of a pipe material. However, since PVC-M samples fail by ductile tearing rather than by brittle fracture, only PVC-U samples can be assessed. The size of each PVC-U pipe sample analyzed is given in Table 6.1.

Using the SENB sample geometry in Figure 6.6, rectangular bar specimens were machined from the pipe wall with a thickness (*B*) and width (*W*) conforming to the requirements of ASTM D5045. As stated previously, this requirement enforces plane-strain conditions and ensures that LEFM theory applies. Notch depths and static loads were chosen to give values of K_I determined from Equation 6.3.

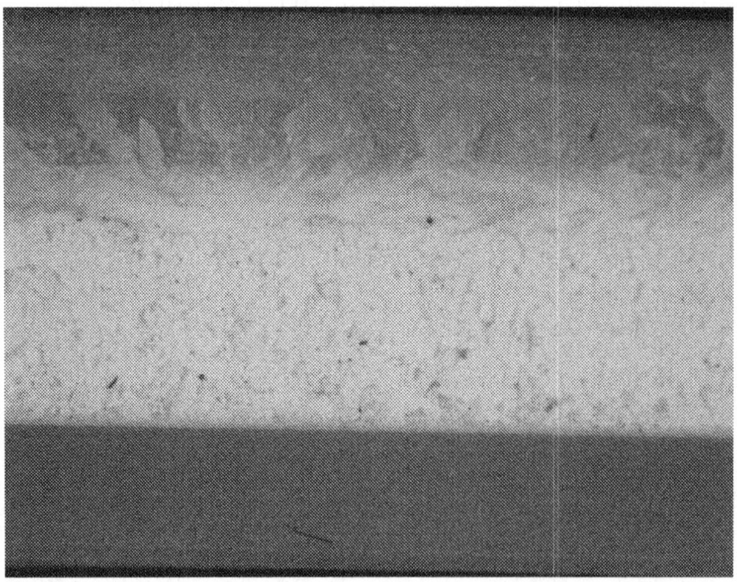

Figure 6.18 Fracture surface from C-ring test on PVC-U Tin (USA)

Crack Incubation

As shown in Figure 6.19, the individual stages of the failure process in an SENB test are reflected in changes in sample compliance over time.

The sample compliance (the ratio of crosshead displacement to load) initially increases sharply, then its rate of increase slows down. Since compliance is the reciprocal of sample stiffness, this change over time indicates creep deformation. This deformation continues until, at some point, an inflexion in the compliance versus time curve is observed, and the sample compliance begins to increase at a constant rate. This indicates the end of Stage I incubation, and the initiation of Stage II crack growth. The sample compliance is now comprised of two components: material creep, and crack growth. Slow crack growth continues until a sharp increase in compliance is observed and the specimen fails by brittle fracture.

Following Bradley, Slay and Self (1998), the total sample compliance can be separated into its individual material creep and crack growth components. The time-dependant material creep compliance, $C(t)$, is estimated by fitting the first part of the total compliance curve with a general power law (Bradley, Slay and Self 1998):

$$C(t) = C_1 + C_2 t^{-r} \quad (6.14)$$

where C_1, C_2, and r are constants obtained from the data fit. Figure 6.20 illustrates the separation of total sample compliance into material and crack growth components.

It follows that the incubation period ends when the total and material compliance curves separate, signaling crack initiation. In this case, a K_I of 3 MPa.m$^{0.5}$ gives an incubation time of t_i = 2.3 hours. Referring back to Equation 6.4, it can be seen that if the δ* required for crack initiation can be determined, it can be related to a crack initiation time (Equation 6.5).

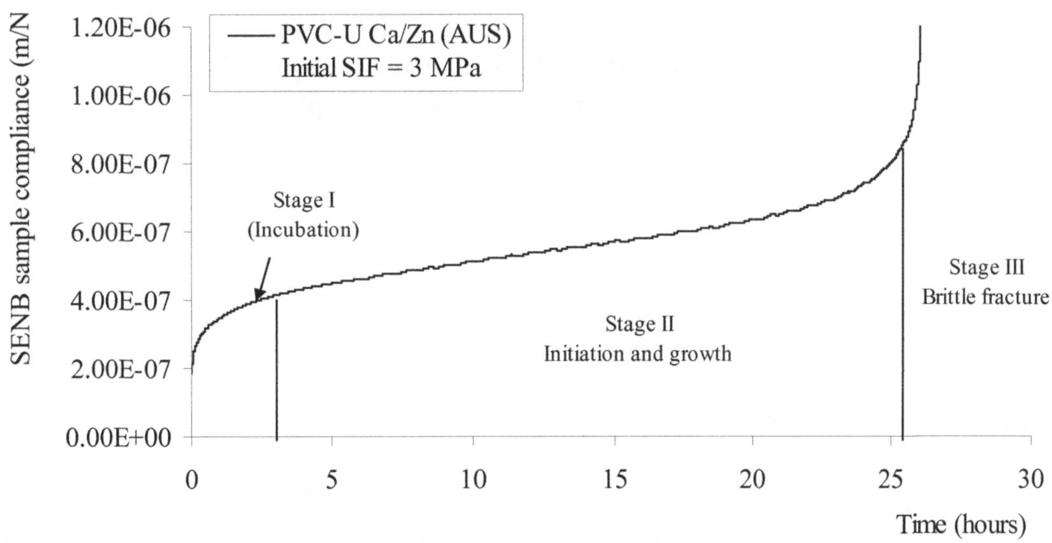

Figure 6.19 Variation in sample compliance with time for SENB test on PVC-U Ca/Zn (AUS) (K_I = 3 MPa m$^{0.5}$)

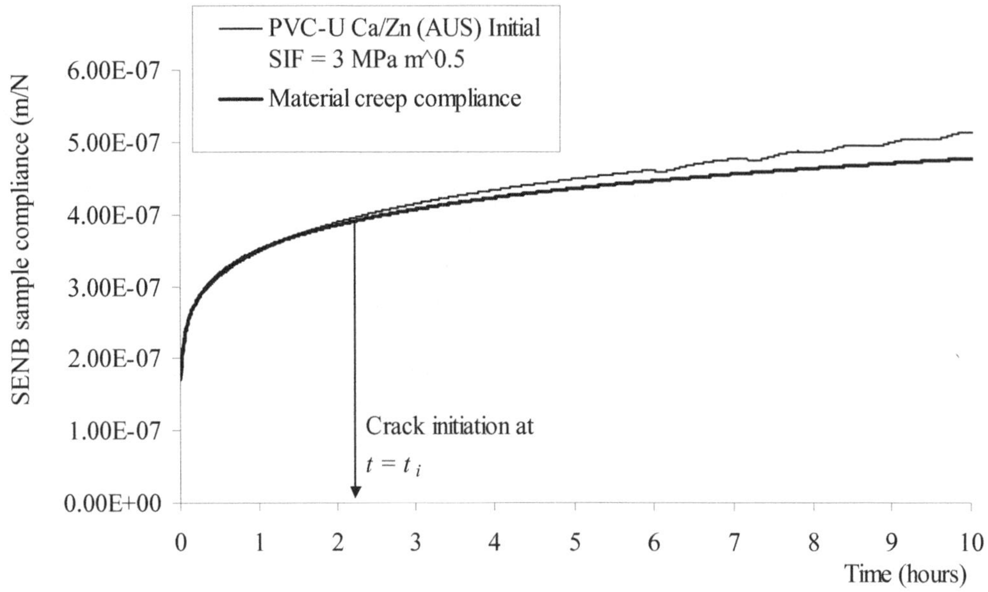

Figure 6.20 SENB testing: separation of total SENB sample compliance into material and crack growth components ($K_I = 3$ MPa m$^{0.5}$ for PVC-U Ca/Zn (AUS))

Table 6.4 summarizes results from SENB tests on PVC-U samples. Data from previous tests on PVC-U imply that when time is expressed in hours, for E_0 = 3,200 MPa, and σ_0 = 45 MPa, the parameters p and q are 0.06 and 0.03 respectively (Davis, Burn, and Whittle 2001)

As shown, PVC-U Tin (USA) exhibits a greater δ^* requirement than the other samples, indicating greater resistance to crack initiation. Referring back to Figure 6.4, this greater resistance to crack initiation partly explains the extended failure times for this sample.

Table 6.4
Critical crack opening displacement required for crack initiation in PVC-U samples

Sample	δ^* (mm)
PVC-U Ca/Zn (AUS)	0.067
PVC-U Tin (USA)	0.088
PVC-U Tin (AUS)	0.072
PVC-U Lead (USA)	0.071

Slow Crack Growth in PVC-U

After incubation and crack initiation, a period of steady, slow crack growth occurs, followed by a period of unsteady growth associated with fast brittle fracture. Referring back to Equation 6.6, the rate of steady crack growth (da/dt) during this period is written as:

$$\frac{da}{dt} = DK_I^m \tag{6.15}$$

where D and m are constants for a particular environment.

Previous researchers (Bradley, Slay and Self 1998) have shown that the crack growth rate in a notched sample can be obtained from the time-dependant changes in compliance. As illustrated in Figure 6.21, by separating material compliance from total compliance, compliance changes due to crack growth alone can be obtained. This change can then be related to changes in crack length via the LEFM compliance relationship (Atkins and Mai 1988). Changes in K_I can be obtained from LEFM theory (Equation 6.3 for an SENB specimen). Note that in this case, K_I refers to the *current* applied SIF (which changes with increasing crack length during an SENB test) and not the initial SIF, K_{Ii}.

If the power-law relationship above is applicable, a linear relationship between log *da*/*dt* and log K_I should be obtained. As shown in Figure 6.22, a reasonably accurate data fit can be achieved.

Based on the current data, the general trend indicates that for the same K_I, the slow crack growth rate in PVC-U Tin (USA) is less than that for the other PVC-U samples. Referring back to Figure 6.4, this increased resistance to crack growth correlates with the longer recorded failure times for a given K_{Ii} in the C-ring test.

Fitting the data for each sample in Figure 6.21 gives values of D and m, as listed in Table 6.5. In each case, K_I is expressed in MPa.m$^{0.5}$ and *da*/*dt* in m/s.

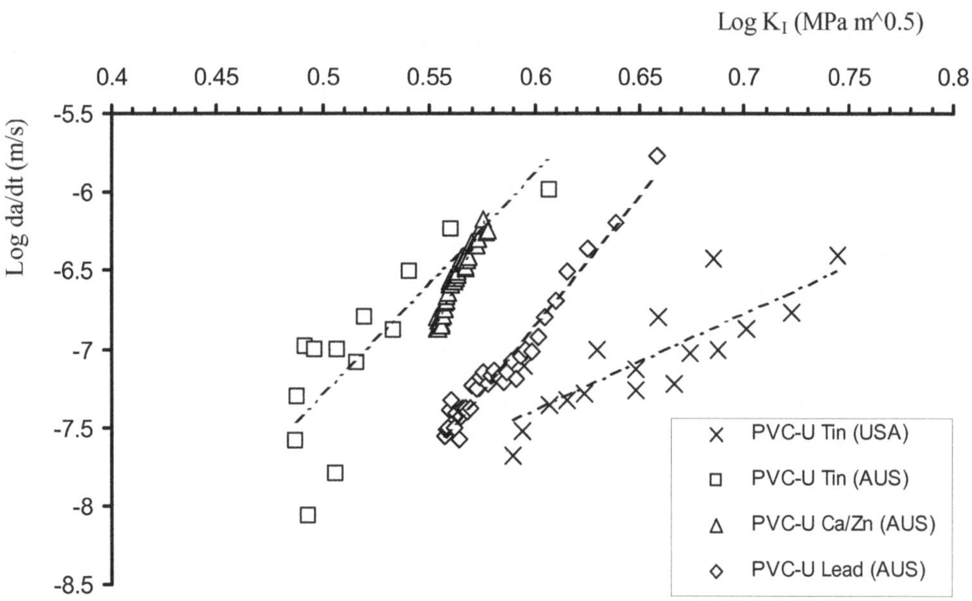

Figure 6.21 SENB testing: variation of crack growth rate with applied SIF for PVC-U samples

Table 6.5
Slow crack growth rate parameters for PVC-U samples

Sample	$Log_{10} D$	Growth rate exponent m
PVC-U Ca/Zn (AUS)	–22.28	27.92
PVC-U Tin (USA)	–11.09	6.16
PVC-U Tin (AUS)	–14.204	13.84
PVC-U Lead (AUS)	–16.65	16.32

EWF TEST RESULTS – PLANE-STRESS FRACTURE RESISTANCE

To determine the resistance to ductile fracture, EWF tests were performed on PVC-U and PVC-M samples. In each case, 2-mm thick samples were machined from the middle of the pipe wall. Samples were aligned with the notches perpendicular to the longitudinal axis of the pipe, and the applied load parallel to the longitudinal axis of the pipe. The direction of crack propagation is therefore around the pipe circumference. However, in the absence of preferential molecular alignment in these pipes, this is also representative of the through-thickness direction of the pipe.

Plane-Stress Fracture Resistance of PVC-U Pipes

The variation of the total work of fracture with ligament length for PVC-U samples is plotted in Figure 6.22.

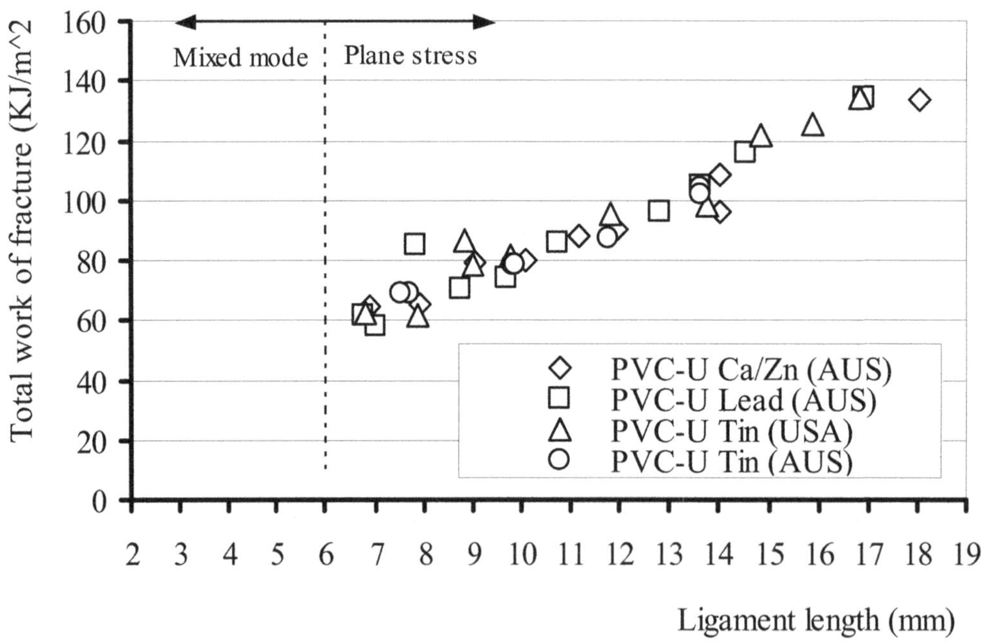

Figure 6.22 EWF tests: variation in total work of fracture with ligament length/specimen thickness ratio for PVC-U samples

As outlined previously, valid EWF tests must be under plane-stress conditions, and yielding must occur across the entire ligament length before specimen separation. To meet this requirement, the ratio between specimen ligament length and thickness must be greater than 3 (Clutton 2001). For ratios below this value, a mixed plane-strain/plane-stress state occurs, and results are difficult to interpret. Examination of results for individual samples indicates that the total work of fracture decreases at a higher rate with decreasing ligament thickness below the plane-stress transition. This implies that the plane-stress ductile component of fracture is also decreasing, and the failure mode is progressing towards plane-strain brittle fracture. While a plane-strain fracture resistance value can be derived from tests in this regime (Saleemi and Nairn 1990), relatively small ligament lengths can reduce the accuracy of recorded loads, and hence final test results.

Referring back to Equation 6.7, EWF results in the plane-stress regime ($l/t > 3$) can be analyzed by fitting a straight line to the data in the form $w_f = w_e + \beta w_p l$. The intercept of this equation gives the plane-stress EWF (or plane-stress ductile fracture resistance, w_e). The gradient of the equation corresponds to βw_p, and is a measure of the non-essential plastic work of fracture associated with deformation remote from the crack tip (Figure 6.7). Table 6.6 summarizes these properties as obtained from the data in Figure 6.22.

As shown, there is little difference in the βw_p parameter for each sample, suggesting that the amount of plastic deformation in the outer plastic zone is similar. However, w_e differs for each sample. PVC-U Tin (AUS) exhibits the highest w_e value at 24.66 kJ/m^2, followed by PVC-U Ca/Zn (AUS) at 19.52 kJ/m^2, PVC-U Lead (AUS) at 15.47 kJ/m^2, and PVC-U Tin (USA) at 14.06 kJ/m^2. While the w_e values are higher than those reported previously for PVC-U pipe (Lau and Truss 2002), it should be noted that w_e is influenced by the level of processing and molecular orientation that is set up in the extruded pipe wall (Kwon and Truss 2002, Lau and Truss 2002). For example, Lau and Truss reported a variation in w_e from 3.0 to 14 KJ/m^2 from the center of the wall to the outer surface of a pipe of nominal diameter 150 mm and wall thickness 14.5 mm (Lau and Truss 2002). The results from the current study would be expected to be higher than those reported previously, since samples in the current study were taken from pipes with thinner wall thickness. Increased shear stress at the center of the thinner pipe wall during extrusion would be expected to promote both higher processing and axial molecular alignment levels. However, further testing would be required to validate such a dramatic influence of pipe processing on w_e.

It is interesting to note that the ranking of PVC-U materials differs to that obtained from C-ring fracture toughness tests, which use samples with the full pipe wall thickness. This would suggest that, in this case, the EWF test is not a useful indicator of resistance to fracture failure in the field. However, further testing is required to reduce the expected variation in results. For example, as outlined by Clutton (2001), a minimum of 20 data points were required to give a 10% confidence interval in w_e for linear low-density polyethylene. While it would be worthwhile to carry out the full assessment of EWF tests on PVC-U to ensure the accuracy of the result, this was not done in this case. However, the EWF test does show promise for assessing thin samples for ductility, but within the scope of this work it is considered that the C-ring test is more applicable for polymers such as PVC-U that fail in a brittle mode.

Table 6.6
Results from EWF tests on PVC-U samples: plane-stress fracture resistance w_e

Sample	w_e (KJ/m^2)	βw_p (KJ/m^3)
PVC-U Ca/Zn (AUS)	19.52	6.12
PVC-U Lead (AUS)	15.47	6.74
PVC-U Tin (USA)	14.06	6.98
PVC-U Tin (AUS)	24.66	5.63

Plane-Stress Fracture Resistance of PVC-M Pipes

As with PVC-U, PVC-M samples were aligned with the notches perpendicular to the longitudinal axis of the pipe and the applied load parallel to the longitudinal axis of the pipe. The direction of crack propagation in EWF tests was therefore around the pipe circumference. For comparison with Figure 6.23, Figure 6.24 shows the variation in total work of fracture with ligament length for PVC-M samples.

Table 6.7 summarizes the EWF properties obtained from the data in Figure 6.22. In comparison to PVC-U, significant increases are observed in w_e for the PVC-M pipes shown in Table 6.7. This indicates that a greater amount of energy is absorbed in producing ductile fracture in the plane-stress regime. This may reflect the influence of impact modifiers used in these materials to promote ductility. Again referring back to the C-ring fracture surfaces, the affinity of PVC-M for ductile tearing rather than brittle fracture is responsible for this increased w_e. Similar to PVC-U, the ranking of PVC-M samples differs from that observed in the C-ring test. However, further tests are required to reduce variation in w_e, as discussed by (Clutton 2001), before the differences in w_e associated with the different formulations can be validated. However, on first appearances it would appear that the Ca/Zn-stabilized pipe was the most ductile pipe.

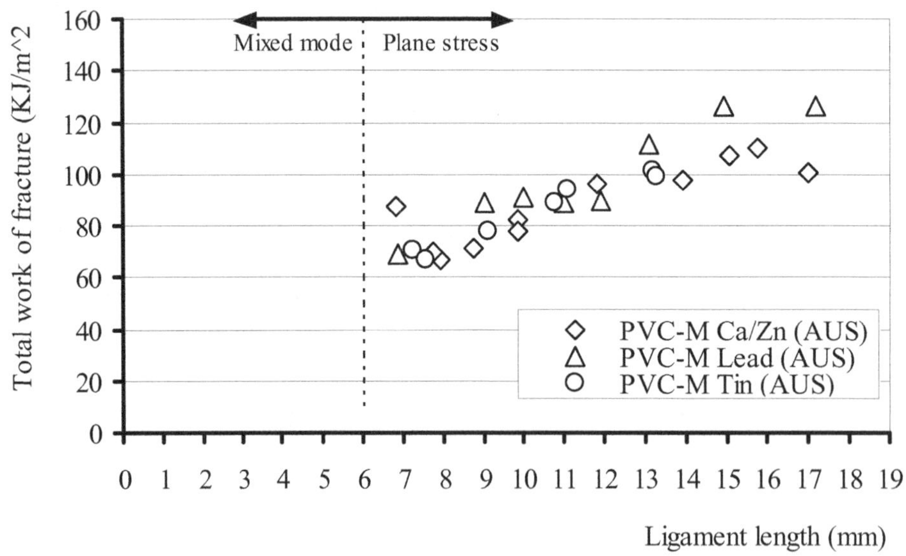

Figure 6.23 EWF tests: variation in total work of fracture with ligament length/specimen thickness ratio for PVC-M samples

Table 6.7
Results from EWF tests on PVC-M samples: plane stress fracture resistance w_e

Sample	w_e (KJ/m^2)	βw_p (KJ/m^3)
PVC-M Ca/Zn (AUS)	45.91	3.73
PVC-M Lead (AUS)	28.48	5.53
PVC-M Tin (AUS)	30.35	5.85

Plane-Stress Fracture Resistance of PVC-O Pipes

Since preferential molecular alignment is produced in the circumferential direction during the manufacture of PVC-O pipe (Chapman and Ågren 1998), it is informative to compare w_e in the longitudinal, radial, and circumferential directions of the pipe. Figure 6.24 schematically illustrates these possible orientations of EWF test samples in PVC-O pipes.

Since the wall thickness of PVC-O pipe samples was too small to obtain samples with radial crack growth, only samples with axial and circumferential crack growth directions were assessed. However, it was reasoned that results from axial crack growth samples should also reflect those from radial crack growth samples; the direction of crack growth is *across* molecular alignment in both cases.

Figure 6.26 illustrates the variation in total work of fracture with ligament length for PVC-O samples.

As shown, results for PVC-O Ca/Zn (AUS) and PVC-O Tin (USA) are similar when samples are oriented with circumferential crack growth. However, a clear increase in total work of fracture is observed when PVC-O Ca/Zn (AUS) samples are oriented with axial crack growth. Table 6.8 summarizes results from Figure 6.26.

As shown, an order of magnitude increase in w_e is observed when PVC-O Ca/Zn (AUS) samples are oriented for axial crack growth. This difference is comparable to results reported by other researchers (Kwon and Truss 2002). Referring back to Figure 6.25, this axial crack growth direction is *across* the direction of molecular alignment. Crack propagation (and the creation of new surface area) in this sample orientation requires molecular chain scission to occur and, consequently, a greater amount of external work. In contrast, circumferential crack growth is *along* the direction of molecular alignment. Consequently, the amount of work required is reduced, since only secondary bonds between molecules must be separated.

This difference in w_e with respect to molecular alignment direction is reflected in C-ring tests conducted on PVC-O samples. The photograph in Figure 6.26 illustrates the direction of crack propagation at sample failure. In contrast to C-ring tests on PVC-U samples (in which radial crack growth is observed), crack growth in this case occurs in the circumferential direction. For a given initial applied SIF, a crack will propagate in the direction of least resistance. For PVC-O, this is the circumferential direction, characterized by a reduced w_e value (Table 6.8).

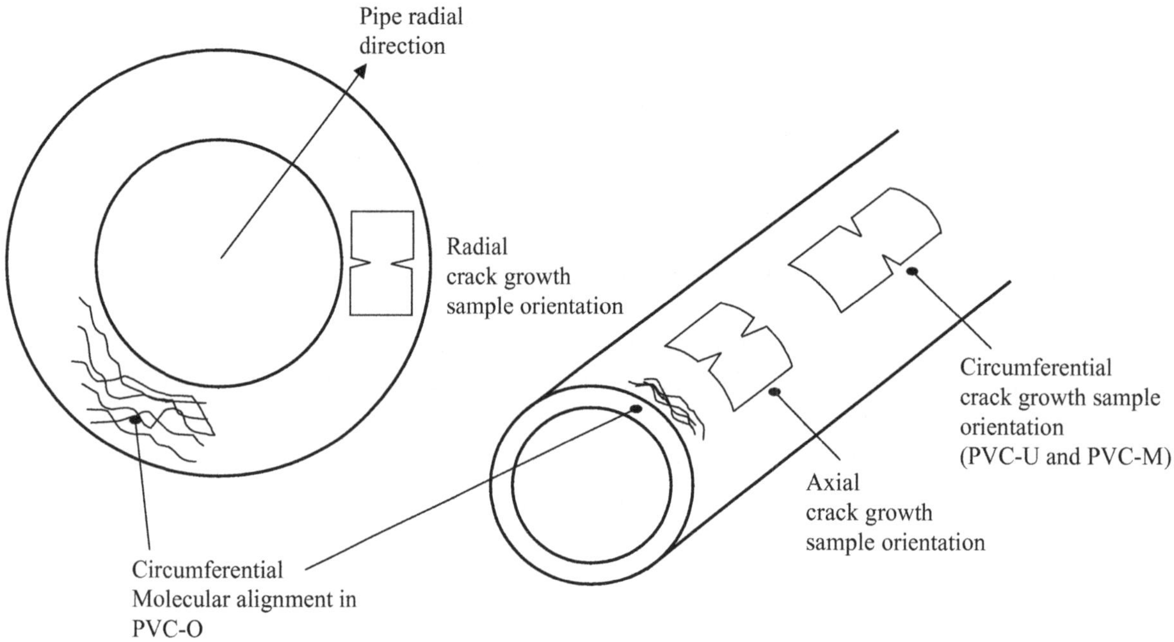

Figure 6.24 Radial, axial, and circumferential crack growth directions for EWF test samples in PVC-O pipe

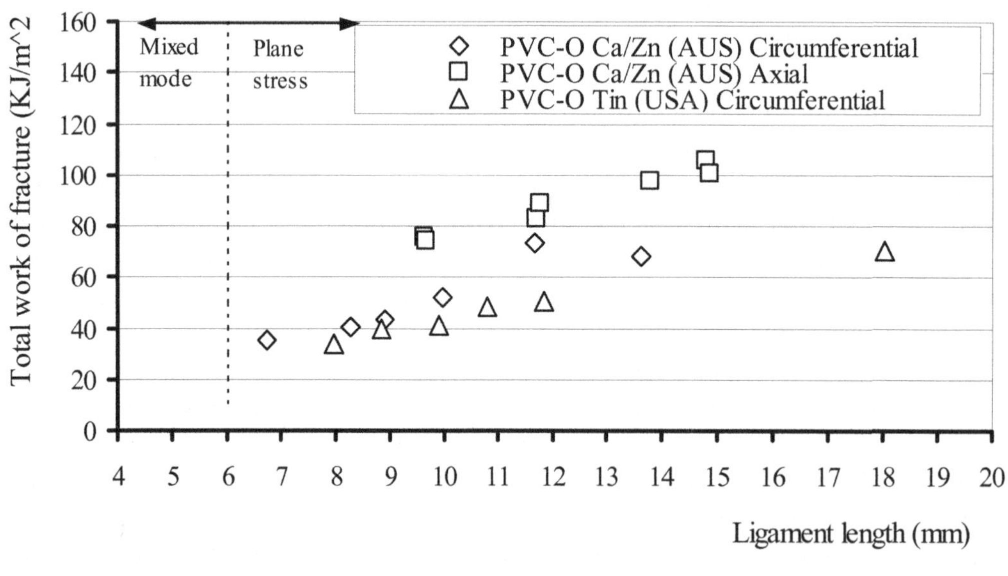

Figure 6.25 EWF tests: variation in total work of fracture with ligament length/thickness ratio for PVC-O samples

Table 6.8
Results from EWF tests on PVC-O samples: plane-stress fracture resistance w_e

Sample	w_e (KJ/m^2)	βw_p (KJ/m^3)
PVC-O Ca/Zn (AUS) (circumferential)	–5.16 (assigned zero)	5.82
PVC-O Tin (USA) (circumferential)	7.49	3.56
PVC-O Ca/Zn (AUS) (axial)	21.00	5.54

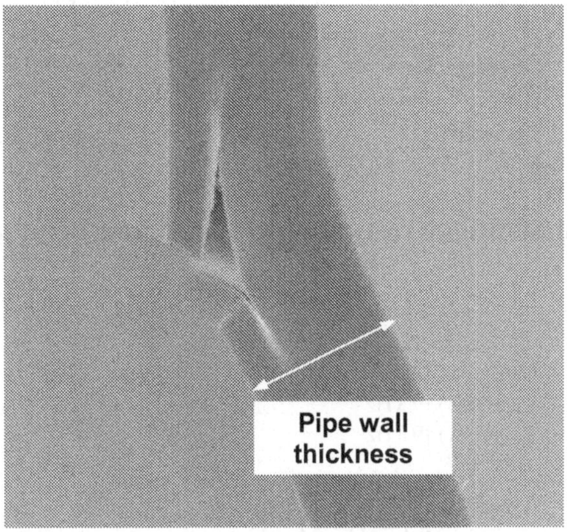

Figure 6.26 Circumferential cracking in C-ring tests on PVC-O Tin (USA) pipe

Fatigue Resistance of PVC-U pipes

Figures 6.27 and 6.28 below compares the fatigue behavior of different PVC samples tested using the rotational bending method. In Figure 6.27, PVC-M Ca/Zn (AUS), PVC-U Ca/Zn (AUS), PVC-U Tin (USA) and PVC-O Tin (USA) are compared at a test frequency of 3.1 Hz.

As reported by Whittle (2002), the performance of PVC-U and PVC-M Ca/Zn (AUS) is similar, with only a slight difference at lower SIF ranges. In a comparative study conducted by Whittle (2002), examination of fracture surfaces from rotational bending tests on PVC-U and PVC-M samples showed similar features, with regions of ductile slow crack growth prior to a smooth region corresponding to final brittle fracture. This suggests that, in contrast to fracture under static loading conditions (in the C Ring test for example), the effect of ductility-promoting modifiers in PVC-M materials on failure mode is not observed. Similar to the C-ring fracture toughness test results, PVC-U Tin (USA) is again superior. Between the SIF range of $\Delta K = 0.7$ MPa m$^{0.5}$ to $\Delta K = 2$ MPa m$^{0.5}$, the number of cycles required for failure is consistently greater for PVC-U Tin (USA) than PVC-U Ca/Zn (AUS). Figure 6.28 compares results obtained at a test frequency of 19.3 Hz.

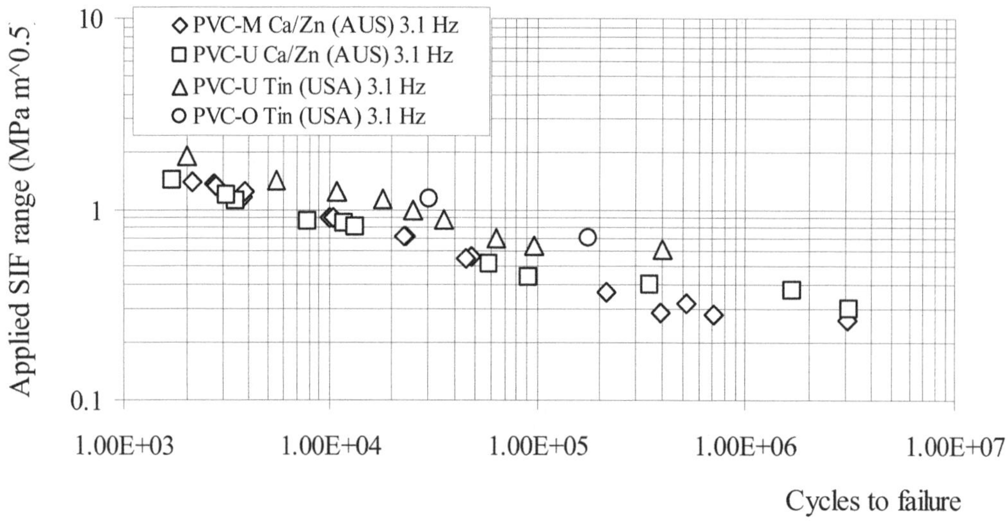

Figure 6.27 Comparison of fatigue resistance in Rotational Bending test at 3.1 Hz

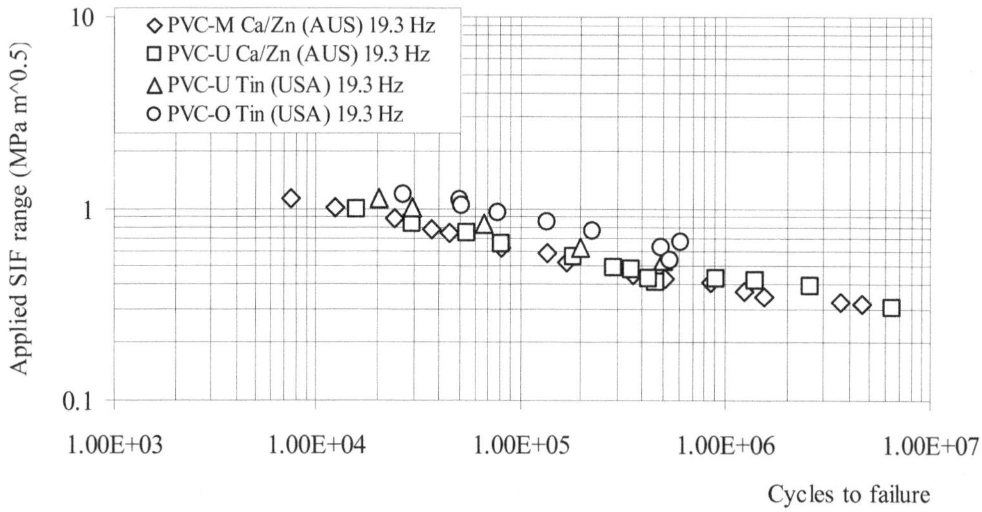

Figure 6.28 Comparison of fatigue resistance in Rotational Bending test at 19.3 Hz

Again, PVC-M and PVC-U exhibit similar fatigue performance, which is in agreement with previous results reported by Whittle (2002). While PVC-U Tin (USA) exhibits a slightly higher fatigue resistance, the difference is less clear. It is interesting to note that PVC-O Tin (USA) exhibits the highest fatigue resistance of all materials tested. While Dukes (1985) reports that "the dynamic fatigue performance of PVC-O pipe "is always better than extruded pipe, when results are obtained by the same means", samples for the rotational bending test are obtained from the sections of the pipe wall that have been pressed flat at elevated temperatures to relieve residual stresses from pipe manufacture. A uniform cross-sectional area in cylindrical rods machined directly from the pipe wall since residual stresses are relived during the machining process. Consequently, the preferential molecular alignment in PVC-O pipes will have been removed during pressing of pipe wall sections at elevated temperatures. However, the increased number of cycles required for failure under a given applied SIF range may indicate the influence of remaining orientation.

CONCLUSIONS & RECOMMENDATIONS

The C-ring test was used to determine the variation in time to failure with initial K_I. For PVC-U samples, tin-stabilized pipe manufactured in the USA exhibited the longest failure times for a given initial K_I; this was followed by Ca/Zn-stabilized pipe from Australia, lead-stabilized pipe from Australia, and finally tin-stabilized pipe from Australia. However, when the C-ring data were extrapolated back to estimate plane-strain fracture toughness values (at a failure time of 15 minutes), all formulations passed the performance requirement stipulated in the AS/NZS 1477. The general trend for PVC-M compared to PVC-U was an increase in failure time for a given initial K_I. However, examination of fracture surfaces of PVC-M C-ring samples revealed that failure occurred by ductile tearing rather than brittle fracture. Consequently, the use of an applied SIF to characterize failure for ductile pipes such as PVC-M is questionable.

The superior performance of PVC-U Tin (USA) pipe in the C-ring test was attributed to an increase in crack incubation times, and a decrease in the slow crack growth rate. Evidence of this was obtained from comparative tests using SENB samples. By extracting material compliance from total sample compliance, the change in compliance due to crack growth alone was determined. Longer incubation times were attained in the SENB test for PVC-U Tin (USA) samples compared to other pipe types. This higher resistance to crack initiation was also reflected in an increased δ^*. It was reasoned that δ^* is a geometry-independent criterion for crack initiation, and therefore partly explains the superior performance of PVC-U Tin (USA) samples in the C-ring test. SENB tests were also used to derive relationships between K_I and the slow crack growth rate. Again, PVC-U Tin (USA) was superior, exhibiting a lower crack growth rate for a given K_I. Since K_I accounts for different geometries and loading types, it was reasoned that this superior crack growth resistance in the SENB test was also present in the C-ring tests on PVC-U Tin (USA). It should be noted that only one pipe manufacturer's material was assessed in this test, and the performance of this pipe may not be indicative of the performance of all manufacturers. Variations in performance are to be expected, thus highlighting the need for suitable performance requirements to be included in the manufacturing standard.

To quantify resistance to ductile fracture, EWF tests were conducted on PVC-U, PVC-M, and PVC-O samples, where w_e was used to characterize ductile fracture. For PVC-U, tin-stabilized pipe manufactured in Australia exhibited the highest w_e value. This was followed by Ca/Zn-stabilized Australian pipe, lead-stabilized Australian pipe, and tin-stabilized pipe manufactured in the USA. As expected, significant increases in w_e were observed for PVC-M, which reflect the influence of ductility-promoting modifiers used in these materials. For PVC-M, the Ca/Zn-stabilized product appeared to be the more ductile. The ranking of PVC-U formulations obtained from EWF was different to that obtained from C-ring fracture toughness testing. Further EWF tests are recommended to reduce the inherent variation in the results, with a minimum of 20 data points having been previously recommended to reduce the variation in w_e of linear low-density polyethylene. This test shows promise for assessing the performance of ductile pipe materials such as PVC-M. However, for pipes that fail in a brittle manner in the field such as PVC-U, the C-ring test is considered more applicable.

EWF testing of PVC-O samples has been used to determine w_e relative to the direction of molecular orientation. Since the test method uses thin samples, it is possible to machine them from the pipe wall in both axial and circumferential directions. This allows the crack growth direction in the test to be aligned either parallel or perpendicular to molecular alignment. Results indicate that the w_e perpendicular to this alignment is much greater than the resistance parallel to alignment. This is explained by the fact that crack propagation (and the creation of new surface area) perpendicular to molecular alignment requires molecular chain scission to occur and, consequently, a greater amount of external work. In contrast, crack growth parallel to the direction of molecular alignment requires only the separation of secondary bonds between molecules. This difference in w_e is reflected in C-ring tests on PVC-O, where crack propagation occurs in the circumferential direction rather than in the radial direction. It is proposed that this is due to molecular alignment in the circumferential direction, and the corresponding increase in crack growth resistance perpendicular to this alignment. These results suggest that, although a relatively low number of samples were tested, the EWF test distinguished between PVC-U and PVC-M, and also between different directions of molecular orientation in PVC-O pipes. However, further data (in excess of 20 data points for example) is required to determine if the test method can distinguish between different PVC-U formulations.

To compare fatigue resistance, rotational bending tests were conducted on notched rod samples. As with the previous fracture tests, PVC-U Tin (USA) exhibited a greater fatigue resistance, consistently requiring a greater number of cycles for failure for a given applied cyclic SIF. Tests were also conducted on PVC-O Tin (USA) samples and indicated that, similar fracture under static load, the resistance of PVC-O to fatigue crack growth was also increased.

CHAPTER 7
DEVELOPING A FRACTURE MECHANICS FAILURE MODEL FOR PVC PIPES

The previous chapters of this report focused on the relevant fracture properties of PVC pipe materials, and on assessing failed pipes, to compile a database of typical defect sizes and damage levels that can lead to fracture failure of PVC pipes in the field. This chapter details how we can use this information to develop a fracture mechanics failure model for a PVC pipe subjected to in-service loading conditions. Since the majority of field failures occur in PVC-U pipes, this chapter focuses solely on this material. As discussed earlier, other failure types, such as those from misplaced rubber rings, are the result of poor installation practices and this needs to be addressed by the industry, as these types of failures cannot be predicted.

APPLIED STRESS INTENSITY FACTOR FOR A BURIED PIPE IN SERVICE

While examination of fracture surfaces suggests that a range of flaw geometries can occur in practice, here we consider the hypothetical case of a radial flaw at the pipe inside surface. As a first approximation, dynamic events such as pressure surges and external impact are ignored. The expected operating conditions of the pipe are then quantified by a total applied SIF, which is composed of individual contributions from internal pressure, external soil loading, and residual stress from the manufacturing process. The schematic diagram in Figure 7.1 illustrates how these individual contributions combine to act on a defect at the pipe inner surface. Provided that point loads are avoided during installation, this loading configuration represents a worst-case scenario for a buried pipe.

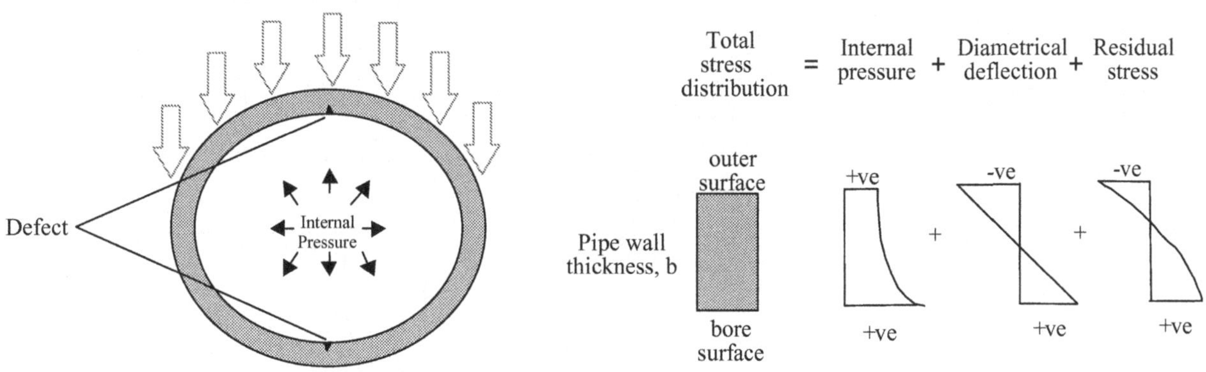

Figure 7.1 Hypothetical pipe geometry containing a defect at the inner surface

The through-thickness stress distributions at the defect location are also shown schematically for each loading component. While the stresses associated with internal pressure and deflection have been well quantified, the stresses associated with manufacturing residual stresses are usually ignored. In most cases, extruded pipes are spray-cooled with water at their outer surface during manufacture. PVC (in common with most polymers) exhibits a low thermal conductivity, which means that this single-surface cooling produces substantially different cooling rates through the pipe wall thickness. Consequently, local density changes and accompanying thermal contractions of molten material vary at different positions in the pipe wall. Material at the outer surface solidifies quickly, preventing the contraction of slower cooling material at the pipe inner surface. This inevitably leads to residual stresses, through the action of frozen-in strains, and the modulus of the cooled material. For residual stresses, the outer layers are in compression, balanced by tension in the inner layers. In the circumferential direction, the shape of this residual stress profile is parabolic, as shown in Figure 7.1 (Williams 1984). Consequently, if residual stresses are added to the stresses associates with deflection and pressure, failure is most likely to initiate at the inner surface of the pipe, due to the high tensile stresses at that point, with any flaw acting as a stress concentrator, increasing the failure potential.

Since each of the stress components shown in Figure 7.1 acts to open an inner surface defect located at the pipe wall, they can be simply added together to give a total applied SIF:

$$K_{IT} = K_{IP} + K_{ID} + K_{IR} \tag{7.1}$$

where K_{IP} is the SIF generated from internal pressure only; K_{ID} is the SIF from external soil deflection loading; and K_{IR} is the SIF generated from the action of residual stresses, processed-in during pipe manufacture.

To determine a general solution for each SIF component of K_{IT}, finite element analysis (FEA) methods are often used (Lu, Davis, and Burn 2003). As shown in Figure 7.2, a two-dimensional FEA model of a pipe cross-section can be used to approximate the case of a pipe with internal defects at the crown and invert.

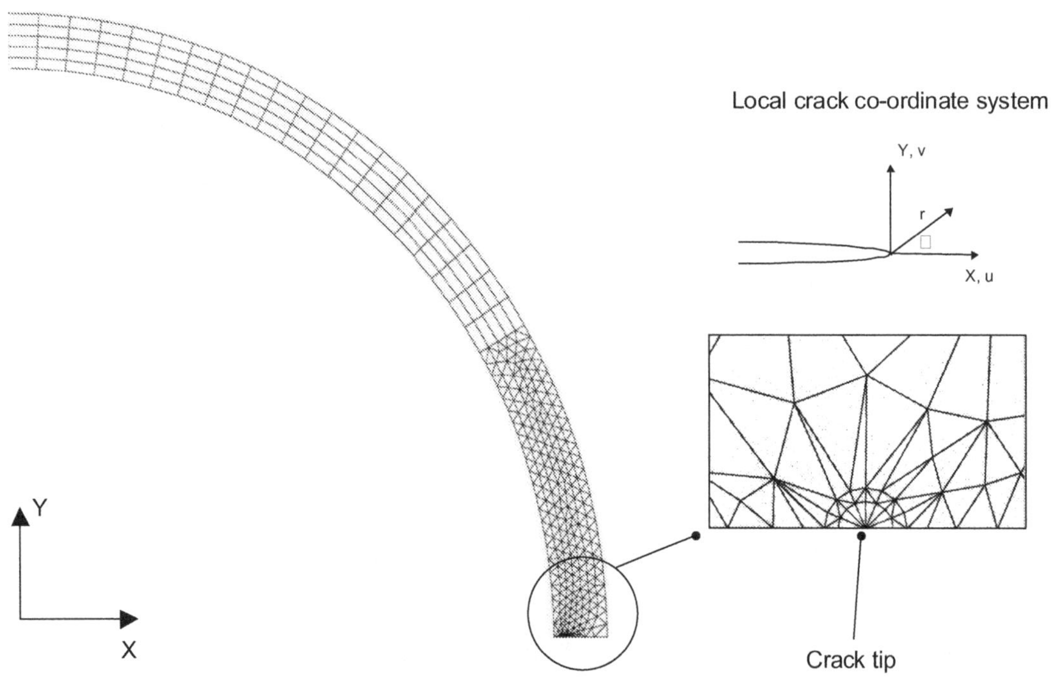

Figure 7.2 Example finite element mesh for pipe with internal defect

Adopting the local crack co-ordinate system shown in Figure 7.2, it can be shown that for a two dimensional model, the variation of an applied SIF K_I along the crack axis is related to normal displacement v, by

$$K_I(r) = \left(\frac{\pi}{2}\right)^{1/2} \frac{E}{2(1-\upsilon^2)} \left(\frac{|v|}{\sqrt{r}}\right) \tag{7.2}$$

where E is Young's modulus, and υ is Poisson's ratio (Paris and Sih 1963). A commercially available FE package (in this case ANSYS 5.6) can then be used to determine these normal displacements and relate them to K_I (ANSYS 1999). Nodal solutions for displacements along a crack front can be determined from an elastic FE stress analysis. These displacements are then normalized along the crack axis r, by setting $v = 0$ at the crack tip. Following Lu, Davis, and Burn (2003), eight-node quadrilateral elements were employed throughout the FE mesh, and singular elements created along the crack front by shifting mid-side nodes to the quarter position (Parks 1979). This ensures that the required square root singularities in stress and strain (stipulated by LEFM theory) are created at the crack tip. Individual SIFs for pressure, deflection, and residual stress are then obtained by applying appropriate loading and boundary conditions to the FE model.

Stress Intensity Factor for Internal Pressure

For a pipe subjected to internal pressure only, the SIF associated with an internal defect is written as (Lu, Davis, and Burn 2003):

$$\frac{K_{IP}}{\sigma_{IP}(\pi a)^{1/2}} = Y_P = 1.824 - 4.669\left(\frac{a}{b}\right) + 14.959\left(\frac{a}{b}\right)^2 - 6.341\left(\frac{a}{b}\right)^3 \tag{7.3}$$

where a is the radial depth of a defect from the pipe inner surface, and b is the pipe wall thickness. Both a and b are measured in metres. Following LEFM notation, Y_P is the geometric correction factor for this geometry and loading type. σ_{IP} is the hoop stress at the pipe inner surface, given by the Lamé equation for a thick-walled cylinder:

$$\sigma_{IP} = \frac{p(R_i^2 + R_o^2)}{(R_o^2 - R_i^2)} \tag{7.4}$$

Here, p is the internal pressure (in Pascals), and R_i and R_o are the pipe inner and outer radii (in meters) respectively. The shape of the Y_P curve follows reasonably closely to the solution in the literature (Lu, Davis, and Burn 2003), with a maximum discrepancy of 5%. For the purposes of the current study, the accuracy of the FE solution is considered adequate.

Stress Intensity Factor for Pipe Deflection

The effect of soil deflection loading can be simulated by imposing a displacement in the negative x-direction, directly ahead of the internal flaw (see Figure 7.1). This creates a membrane hoop stress similar to that imposed by diametrical deflection under soil loads. The corresponding FE solution for external soil deflection loading is (Lu, Davis, and Burn 2003):

$$\frac{K_{ID}}{\sigma_{ID}(\pi a)^{1/2}} = Y_D = 0.803 + 0.353\left(\frac{a}{b}\right) - 1.640\left(\frac{a}{b}\right)^2 - 1.003\left(\frac{a}{b}\right)^3 \tag{7.5}$$

In this case, σ_{ID} is the remotely applied membrane hoop stress at the location of the defect at the pipe inner surface. Castigliano's strain energy theorem can be used to show that:

$$\sigma_{ID} = \frac{2Eb\Delta}{D_m^2\left[\frac{\pi}{4} - \frac{2}{\pi}\right]} \tag{7.6}$$

where Δ is the diametrical deflection imposed directly above the point of interest (in meters), and D_m is the mean diameter of the pipe (in meters). E is the time-dependant Young's modulus of the pipe material (in Pascals).

Stress Intensity Factor for Residual Stress

As discussed previously, for an extruded pipe that has been spray cooled on its external surface during manufacture, a parabolic through-thickness residual stress profile can be assumed, described by:

$$\sigma_R = \sigma_{IR}\left[1 - 3(r/b)^2\right] \tag{7.7}$$

where r is the radial distance from the pipe inner surface, and σ_{IR} is the magnitude of residual stress at the pipe inner surface (in Pascals) (Williams 1984). By applying this distribution internally across the pipe wall in the FE model (with defect faces remaining free), the SIF associated with residual stress is:

$$\frac{K_{IR}}{\sigma_{IR}(\pi a)^{1/2}} = Y_R = 0.727 + 2.791\left(\frac{a}{b}\right) - 2.214\left(\frac{a}{b}\right)^2 - 2.149\left(\frac{a}{b}\right)^3 \tag{7.8}$$

VALIDATING THE FRACTURE MECHANICS FAILURE MODEL

Having established applied SIFs for internal pressure, external deflection and residual stress loading, they can now be used to predict the time to failure for a pipe containing a radial defect at its inner surface. However, the magnitude of residual hoop stress at the inner surface of a PVC-U pipe must first be determined.

Determining the Residual Stress Profile in PVC-U Pipes

While several methods to determine residual stress profiles are described in the literature, a relatively simple method using a single specimen was used. Following Whittle (2002), the residual hoop stress at the inner surface of PVC-U pipes was determined using a short sample of pipe. The section was cut from the pipe such that its length was equal to the outside diameter of the pipe. A longitudinal strip was then cut from the section. The creation of new free surfaces released the residual hoop stresses in the pipe section, and it attempted to close at the location where the longitudinal strip was removed. By measuring the pipe sample diameter before and one hour after strip removal, the residual hoop stress can be calculated from:

$$\sigma_{IR} = \frac{4E'b(D_1 - D_2)}{3(D_2 - b)(D_1 + D_2 - 2b)} \tag{7.9}$$

where E' is the Young's modulus of the material at one hour, D_1 is the initial outside diameter before strip removal, and D_2 is the outside diameter one hour after strip removal (Whittle 2002).

Table 7.1 summarizes the results from residual stress tests. As shown, similar residual stresses were obtained for PVC-U Ca/Zn (AUS) and PVC-U Tin (USA) pipes. While PVC-U Tin (USA) exhibits a slightly lower value of σ_{IR}, it should be noted that its wall thickness is also lower. Consequently, the difference in cooling rate at the inner and outer surface of the pipe during extrusion will be reduced. In turn, the through-thickness gradient of density and thermal contraction will also be reduced. With this in mind, a slightly lower residual stress would be expected in PVC-U Tin (USA) pipe.

Corrected Slow Crack Growth Parameters for PVC at Low Applied SIFs

While the previous SENB tests are useful to compare different PVC materials, examination of Figure 6.16 indicates that these tests were conducted at relatively high applied SIFs. Consequently, they describe crack growth just prior to brittle fracture. To characterize crack growth resistance more accurately over a range of test times, the previous results from the C-ring tests, conducted under a range of applied SIFs, can be used. According to Atkins and Mai (1988), the fracture toughness of plastic materials decreases over time following a power law:

$$K_{IC}(t) = E_0(\varepsilon_y \delta *)^{0.5} t^{-n} \tag{7.10}$$

where E_0 is the short-term materials Young's modulus, ε_y is the material yield strain, and δ_c is the critical crack opening displacement required for crack growth. The parameter n is the time exponent for fracture toughness. It follows that results from the previous C-ring fracture toughness tests can be fitted to Equation 7.10 to provide estimates for these parameters. Based on the experimental data in Figure 6.10, derived parameters for PVC-U Ca/Zn (AUS) and PVC-U Tin (USA) are given in Table 7.2.

Table 7.1
Residual hoop stress at the inside surface of PVC-U pipes

Sample	Nominal diameter	Outer diameter mm (inches)	Wall thickness mm (inches)	Residual hoop stress at inner surface σ_{IR} (MPa)
PVC-U Ca/Zn (AUS)	DN 100 mm	122.2 (4.81)	8.8 (0.35)	3.56
PVC-U Tin (USA)	DN 4 inches	122.0 (4.80)	7.1 (0.28)	3.16

Table 7.2
Parameters describing the time dependence of fracture toughness in PVC

Sample	$E_0(\varepsilon_y\delta^*)^{0.5}$	n
PVC-U Ca/Zn (AUS)	3.99	0.056
PVC-U Tin (USA)	4.00	0.048

Results obtained from tensile testing under static load suggest that E_0 = 2,700 MPa and ε_y = 1.3% are reasonable estimates for PVC-U (manufacturers' data). Substituting these values in the parameter groups in Table 7.2 provides new estimates of the δ^* required for crack growth (Table 7.3).

While the values of δ^* from the C-ring test are approximately twice those from the SENB test, it is encouraging to note that the differences between materials are maintained. Again, PVC-U Tin (USA) exhibits an increased resistance to crack initiation. Since the previous C-ring tests were conducted over a large range of applied SIFs (compared to narrow range of SIFs applied in the SENB tests), the C-ring-based values of δ^* were used for the development of the fracture failure model.

These parameters can now be used to derive corrected slow crack growth parameters, which are based on the C-ring test. Atkins and Mai (1988) propose that *da*/*dt* can be written as:

$$\frac{da}{dt} = \left(\frac{\pi}{8}\frac{\delta^*}{\varepsilon_y}\right)\left(E_0{}^2\varepsilon_y\delta^*\right)^{-\frac{1}{2n}} K_I{}^{\frac{1}{n}} \tag{7.11}$$

Referring back to Chapter 6, it can be seen that Equation 7.11 follows the same form as the empirical power law equation previously used to describe *da*/*dt* (Equation 6.6):

$$\frac{da}{dt} = DK_I{}^m$$

Therefore, substituting the values in Tables 7.2 and 7.3 into Equation 7.11 gives the new (corrected) slow crack growth rate parameters in Table 7.4.

Again, since a larger range of SIFs were applied in the C-ring test, these corrected parameters were used to develop a fracture failure model for PVC pipes.

Table 7.3
Comparison of critical COD from C-ring and SENB tests

Sample	δ* (based on C-ring) (mm)	δ* (based on SENB) (mm)
PVC-U Ca/Zn (AUS)	0.129	0.067
PVC-U Tin (USA)	0.173	0.088

Table 7.4
Comparison of critical COD from C-ring and SENB tests

Sample	$Log_{10} D$ (C-ring)	$Log_{10} D$ (SENB)	Growth rate exponent *m* (C-ring)	Growth rate exponent *m* (SENB)
PVC-U Ca/Zn (AUS)	–12.14	–22.28	17.80	27.92
PVC-U Tin (USA)	–14.9	–11.09	20.83	6.16

Comparing Experimental and Predicted Failure Times for PVC-U Pipes Under Combined Pressure and Deflection

To calculate the predicted failure times, we need to combine the times associated with crack incubation and slow crack growth. By substituting the values determined from the C-ring test into Equation 7.12, we can calculate the crack incubation time, and by substituting into Equation 7.13, we can calculate slow crack growth time:

$$t_i = \left[\frac{E_0 \sigma_0 \delta^*}{K_{IT}^2} \right]^{\frac{1}{p+q}} \tag{7.12}$$

$$\frac{da}{dt} = DK_{IT}^m \tag{7.13}$$

The parameters required to implement Equations 7.12 and 7.13 have been determined previously in Tables 7.3 and 7.4. For known values of internal pressure, diametrical deflection, residual hoop stress, and internal notch depth, Equations 7.3–7.8 can be used to determine K_{IT}. Incubation and crack growth can then be simulated over time until brittle fracture occurs, when K_{IT} exceeds the material fracture toughness K_{IC}.

To simulate incremental crack growth, Equation 7.13 is discretized into a sequence of small time steps, each of duration Δt. If these time steps are of sufficiently small duration, then Equation 7.13 can be explicitly rewritten as:

$$a_{new} = a_{old} + \Delta t D K_{IT}^m \tag{7.14}$$

At any instant in time, the current values of a_{old} and K_{IT} can be used to calculate the new crack length at the end of a time step. Thus, Equation 7.14 can be used in a time-marching loop to incrementally grow a crack until K_I exceeds the material fracture toughness.

To validate the fracture model, a series of combined pressure/deflection tests were conducted on DN 4 inch, Class 250 psi, PVC-U Tin (USA) pipes, as shown in Figures 7.3 and 7.4.

A 775-mm (30.5-inch) long section of pipe with an axial notch of 150 mm was sealed at each end with external end caps and supported by a central plate, as shown. The pipe cross-section was deflected by a spring-loaded plate, which acted over a length of 400 mm. While pipe sections where subjected to an initial deflection to simulate installation conditions, the true deflection for the beginning of each test was recorded after the pipe cross-sections partially re-rounded under internal pressure.

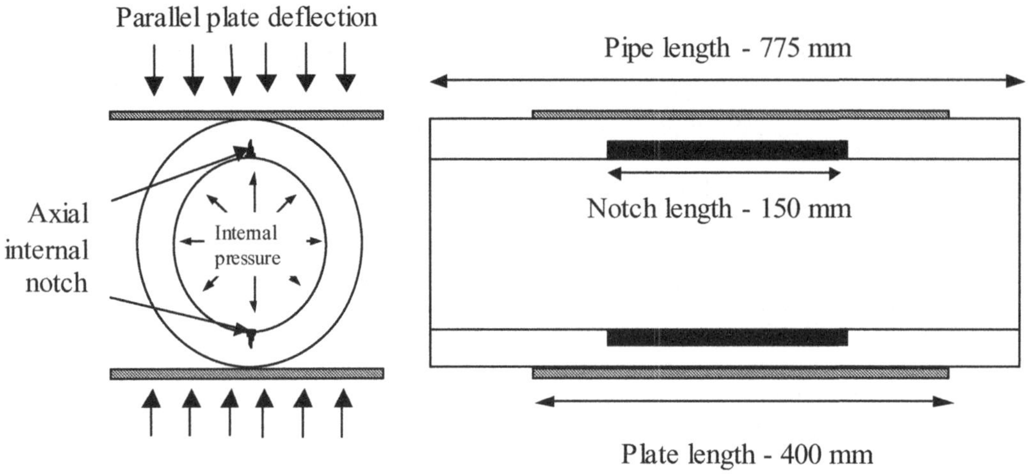

Figure 7.3 Experimental pressure and deflection loading arrangement

Figure 7.4 Photograph of experimental pressure and deflection loading rig

Figure 7.4 shows the experimental arrangement. Test conditions and results are given in Table 7.5. Figure 7.5 shows the brittle fracture failure of a pipe. This failure mode was observed in all combined pressure/deflection tests conducted, While Table 7.6 lists the incubation, slow crack growth, and total predicted failure times from the fracture failure model. As shown, there is reasonable agreement between recorded and predicted failure times across the range of applied loading conditions. Having validated the fracture failure model under loading conditions close to those observed in the field, it can now be developed further, as discussed in the next chapter, to account for failures in large PVC pipe networks.

Table 7.5
Conditions for combined pressure/deflection testing of DN 4 inch, Class 250 psi, PVC-U Tin (USA) pipe

Sample	OD (mm)	b (mm)	p (MPa)	D (mm)	a (mm)	t_f (hours)
1	114.62	6.75	3.50	4.62	0.40	944.33
2	114.67	6.77	3.80	3.40	0.68	49.25
3	114.62	6.78	3.80	4.23	0.87	3.31
4	114.63	6.73	4.00	5.40	0.59	102.73
5	114.68	6.78	4.50	5.45	0.65	5.53

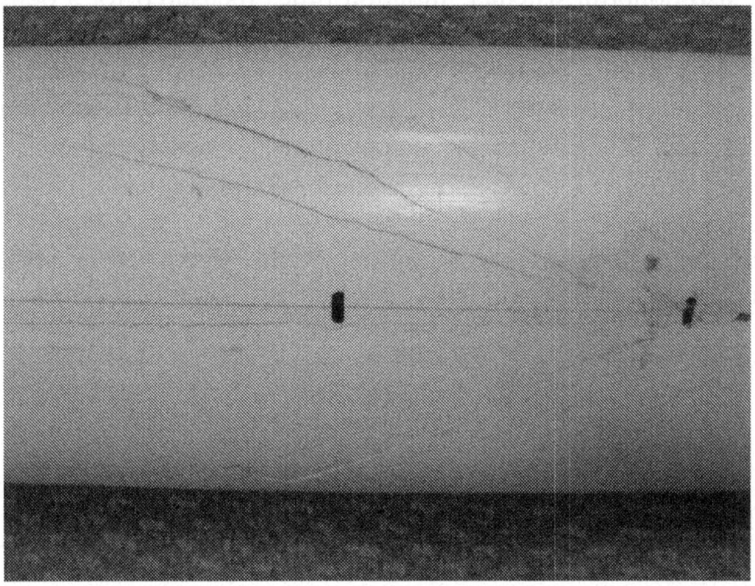

Figure 7.5 Brittle fracture of DN 4 inch, Class 250 psi, PVC-U Tin (USA) pipe after testing under combined pressure and deflection

Table 7.6
Comparison of recorded and predicted failure times in DN 4 inch, Class 250 psi, PVC-U Tin (USA) pipe

Sample	K_{IT} (MPa m $^{0.5}$)	t_f (hours)	t_i (hours)	t_{cg} (hours)	$t_i + t_{cg}$	Error (hours)
1	3.54	944.33	973.39	127.00	1,100.93	+156.0
2	4.04	49.25	52.97	18.00	70.97	+21.72
3	5.00	3.31	0.46	2.00	2.46	–0.85
4	4.87	102.73	0.81	46.00	46.81	–55.92
5	5.39	5.53	0.08	2.00	2.08	–3.45

Incubation Periods for Pipes in Service With Lower Applied SIFs

While the use of incubation times and slow crack growth rates in the laboratory/modeling studies has indicated that the correlation between the two is fairly close, the presence of an incubation time for plastic pipes in service is a subject of some debate. For example, in a series of fatigue tests on razor-notched samples of high-density polyethylene, it has been shown that the estimated incubation times before crack growth is the same order of magnitude as the estimated lifetime during crack growth (Bucknall and Dumpleton 1985). As reported by Truss (1988), similar fatigue experiments on PVC resulted in slightly longer incubation periods than those found for polyethylene. However, it should be noted that (similar to the current study) these previous experiments were conducted over a limited range of K_I. Truss (1988) notes that extrapolation of incubation times from experimental studies at high applied stresses, to the PVC-U pipe at their low operating pressures in service, "must be dubious". This view is reinforced by Joseph and Leevers (1985), who ignored the possible incubation period and still reported good agreement with recorded and predicted failure times for PVC-U under cyclic loads. According to Truss (1988), it is also doubtful whether initiation times measured for a deliberately introduced razor notch are the same as those that would occur from an inherent defect in the pipe wall. Unpublished work by CSIRO has shown similar results to those of Bucknall and Dumpleton (1985), in that the incubation time in PVC-U for a blunt notch was twice that of a sharp notch, however a transition was found when the notch radius was sharper than 5 μm (i.e., a scalpel notch with a radius of 0.1–0.2 μm), and for these sharper notches no incubation time was observed.

To illustrate the problem of low K_I further, Figure 7.6 plots the variation of incubation time with K_I according to Equation 7.12.

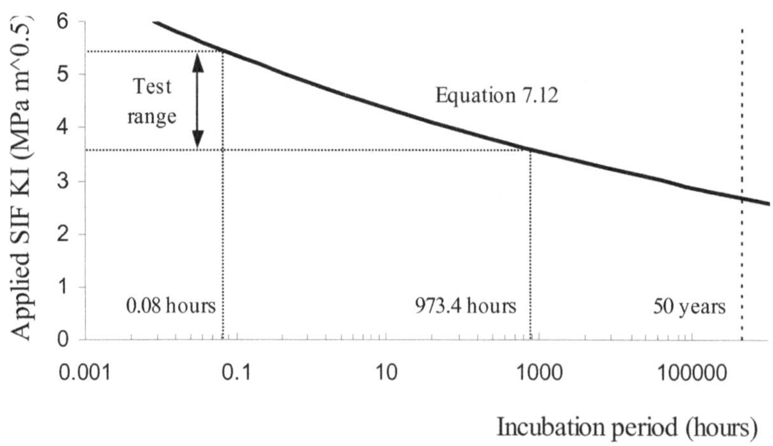

Figure 7.6 Variation of predicted incubation time with initial applied SIF for PVC-U Tin (USA) pipe

As shown, to obtain results in a practical time frame, applied SIFs were restricted to a relatively narrow range between (approximately) 3.5 and 5.5 $MPa.m^{0.5}$. This corresponds to predicted incubation times between 0.08 and 973.4 hours, which (when combined with subsequent slow crack growth periods) gave the predicted total failure times in Table 7.6. For these relatively high initial K_I values, good agreement between predicted and observed crack incubation time has previously been reported (Williams 1978). However, it can also be seen how the predicted incubation times (plotted on a log scale) increase dramatically for lower K_I. For example, an incubation time of 50 years is predicted for $K_I = 2.7$ MPa.m$^{0.5}$.

For PVC-U pipes in service, initial K_I values must be limited to lower values for two reasons. Firstly, from fracture surface examinations of field failures in Chapter 5, inherent defect size is small. Secondly, pressure ratings limit the operating pressures that can be attained in service. As an example, consider the PVC-U DN 100 Class 1.2 MPa pipe that was reported as a longitudinal fracture failure in Victoria, Australia. As shown in Figure 5.11, the fracture surface from the exhumed pipe indicates that crack initiation occurred at an inherent cavity, approximately 1-mm long, in the pipe wall. The outer and inner radii for this pipe were 57.1 and 51.1 mm respectively. If we assume that this pipe was operating at its pressure rating of $p = 1.2$ MPa, then Equation 7.3 gives an applied SIF due to internal pressure of $K_{IP} = 0.88$ MPa.m$^{0.5}$. According to Equation 7.12, this would result in an incubation period of $t_i = 3.1 \times 10^{12}$ years. In practice, the pipe failed after 34 years in service. The discrepancy between these Figures suggests that, While fracture mechanics theory can predict incubation periods under relatively high K_I values (Williams 1978), an alternative method is required to determine incubation for pipes in service, which operate under low K_I. This supports the unpublished work of CSIRO that shows that where the crack initiation site is considered to be a sharp notch (<5-μm radius), initiation times do not exist. If we assume that all pipes contain a flaw that can be considered to be sharp, then the major factor controlling pipe lifetimes is the slow crack growth rate.

ALTERNATIVE FAILURE CRITERION FOR DUCTILE FRACTURE

The linear elastic fracture mechanics approach adopted in this study assumes that failure occurs by brittle fracture, and that any plastic deformation is restricted to small zones ahead of a propagating crack. An examination of failed samples from the combined pressure deflection tests indicates that all failures occurred by brittle fracture. However, in some failures that occur in the field, examination of fracture surfaces reveals evidence of local plastic deformation, and ductility accompanying fracture. While elastic-plastic fracture mechanics theory can be used to develop more complex failure models, a simpler method that has previously proved successful is the net section yield criterion proposed by Marshall, Pearson, and MacKellar (1996). The net section yield stress, σ_{NS}, decreases linearly with increasing crack size according to:

$$\sigma_{NS} = \sigma_Y \left(1 - \frac{a}{b}\right) \tag{7.15}$$

where σ_Y is the time-dependant material yield strength. For a buried pipeline in the field, ductile fracture is assumed to occur when a crack has grown sufficiently long that σ_{NS} has decreased below the total applied stress, σ_{TOT}, which is the sum remote stresses due to internal pressure (σ_{IP}), diametrical (σ_{ID}), and residual stress (σ_{IR}). As shown previously in Equation 7.6, σ_{ID} is described in terms of the diametrical deflection (Δ) imposed on a pipe. In practice, Δ is usually unknown, but can be determined from the magnitude of external loads due to soil backfill, and any live surface load applied. As outlined in AS/NZS 2566.1 (Standards Australia/Standards New Zealand 1998), diametrical deflection in a buried flexible pipe is based upon the Marston Spangler theory from Iowa University and is given by:

$$\Delta = \frac{0.1 \times 10^{-3} D_m \left(\gamma H + P_{surf}\right)}{\frac{8EI}{D_m{}^3} + 0.061 E_{soil}} \tag{7.16}$$

where D_m is the mean pipe diameter (in meters), γ is the unit weight of soil backfill (in kN/m^3), H is soil cover depth (in meters), P_{surf} is the external surface loading (in kPa), E is the Young's modulus of the pipe material (in MPa), and E_{soil} is the soil modulus (in MPa). I is the second moment of area of the pipe wall for ring bending and is given by:

$$I = \frac{b^3}{12} \tag{7.17}$$

CONCLUSIONS AND RECOMMENDATIONS

In this chapter, a fracture mechanics failure model has been developed to predict failure times for PVC pipes under combined internal pressure and deflection loads. The method combines time-dependant fracture toughness, slow crack growth rate, and individual K_I values for internal pressure, diametrical deflection, and residual stress loading. Using previous results (Lu, Davis, and Burn 2003), K_I values for internal pressure, diametrical deflection, and residual stress loading were obtained from FEA. These SIFs were added to define a total applied SIF for in-service loading conditions.

Since the previous slow crack growth tests (using the SENB test) in Chapter 5 were conducted over a relatively small range of K_I, results from the C-ring fracture toughness test in Chapter 5 were reanalyzed. Values of δ^*, D, and m, were obtained from the variation of K_I with recorded failure time in the C-ring test. As with the SENB test, the increased resistance of PVC-U Tin (USA) pipe to crack initiation and growth was again observed.

The fracture mechanics failure model was compared against experimental simulation of in-service loading, in which an internally notched pipe was pressurized and subjected to parallel plate deflection loads. The fracture mechanics model was used to predict an incubation period (before δ^* was obtained), and a period of slow crack growth to failure. Failure was predicted to occur at a critical crack length when K_I exceeded the material fracture toughness. The model predictions agree well with recorded failure times under the range of experimentally applied SIFs. While this agreement is encouraging, the applicability of the incubation period for pipes in service requires further validation, as there is evidence that they may not exist where sharp notches exist in the pipe wall.

CHAPTER 8
EXTRAPOLATING FRACTURE MECHANICS FAILURE OF INDIVIDUAL PIPES TO PIPE NETWORKS

While modeling of individual pipe assets can provide information on the performance of those assets, water utilities normally wish to try and predict the performance of their network. To do this, we could incorporate the characteristics for all assets in the network, and calculate lifetimes based on the models developed in Chapter 7. This would need us to define and apply values for each of the relevant factors in the equations, such as external load, damage function, and so on to each asset in the network. Obviously, much of this data is not available and valid estimations of these values based on uncertainty need to be made.

In order to extrapolate failure models from individual pipes to pipe networks, uncertainty in key variables must be included. The first step in such a probabilistic failure analysis is to formulate a "limit-state" failure criterion in terms of the load (L) and resistance (R) variables

$$m = R\text{-}L \tag{8.1}$$

where failure is defined to occur when $m < 0$. As defined previously, the criterion for brittle fracture is when the total applied stress intensity factor exceeds the material fracture toughness:

$$m_1 = K_{IC} - K_{ITOT} \tag{8.2}$$

where brittle fracture is defined to occur when $m_1 < 0$. As outlined in Chapter 7, a simple, alternative failure criterion for ductile fracture is the difference between σ_{NS} and σ_{TOT} from operating loads (or $m_2 < 0$ for failure):

$$m_2 = \sigma_{NS} - \sigma_{TOT} \tag{8.3}$$

where failure occurs when $m_2 < 0$. Although techniques exist to develop probabilistic versions of these failure criteria, they often rely on simplifying assumptions (Ahammed and Melchers 1997). For example, Level II first-order second-moment (FOSM) methods are often used to account for uncertainty in limit-state models (Melchers 1999, Thoft–Cristensen and Baker 1982). In all failure models that account for uncertainty, the basic variables that comprise the limit-state criteria (i.e., defect size, operating loads, etc.) are no longer single valued, but are treated as stochastic. This means their values follow probability distributions, which have mean values and variances associated with them. The limit-state failure criteria m_1 and m_2 are reformulated in the form $M = f(\overline{Z})$, where M is the stochastic limit-state variable, and $\overline{Z}$ is a random vector of the stochastic variables that comprise the limit-state failure criteria (i.e., defect size, operating loads, etc.).

In FOSM methods, M is assumed to follow the normal probability distribution. Because of this assumption, there is a direct translation between the probability of failure (the probability that $M < 0$) and the shortest "distance" from the failure region defined in variable space and the origin (Melchers 1999). In reality, however, stochastic limit-state variables may not be normally distributed. Complex loading conditions also mean that limit-state failure models are rarely linear, and M is often a non-linear function of stochastic variables. This causes some degree of non-normality in M. Furthermore, iterative methods to calculate β require linear approximations to the failure surface, resulting in oversimplification. As a consequence of all assumptions, realistic physical failure mechanisms can be lost in order to obtain a solution (Richardson 2002).

A more straightforward way to estimate the lifetime probability functions (discrete probability density functions) of PVC pipes is to use Monte Carlo simulation in conjunction with the limit-state formulations of pipe failure. Monte Carlo simulations sample lifetimes by repeatedly generating random numbers, and using these for calculating and evaluating the limit-state equations. For a simulated hypothetical network of pipes, a number of failures are recorded over time, which allows the lifetime distribution to be estimated. In contrast to FOSM methods, Monte Carlo simulations allow the details of the physical failure mechanism to be preserved. It is a bottom-up computer simulation approach that allows for weak and strong exploration (Richardson 2002). In essence, it allows for both quantitative (numerical sensitivity analysis) and qualitative analysis (testing assumptions versus model). Changing the emphasis of the Monte Carlo simulation is easy, simply being a matter of code change. This reverts the modeling focus to the realism of the physical model, leading to increased understanding of the underlying process.

MODELING UNCERTAINTY IN STOCHASTIC VARIABLES

To implement a Monte Carlo simulation, stochastic variables that influence the limit-state failure criteria must be identified. For a buried PVC pipe in service, possible stochastic variables are:

- Pipe size.
- Fracture properties of pipe material.
- In-service loading conditions.
- Inherent defect size

It is reasonable to assume that in order to meet performance standard requirements, pipe size is well controlled during manufacture. Variation in pipe wall thickness and diameter can therefore be assumed to be insignificant. Although in-service loading conditions are uncertain, sufficient data to describe this uncertainty is currently available from water utilities. For plastic pipes, fracture failures are most sensitive to changes in inherent defect size (Barker, Bowman and Bevis 1983), and (to a lesser extent) the parameters that describe slow crack growth resistance (Lu, Davis, and Burn 2003). In the absence of data describing the variation in slow crack growth resistance for a particular PVC material, only variation in inherent defect size is considered in this study. As a useful first approximation, results from the SEM study of field failures conducted in Chapter 5 as well as other data available in the literature (Burn 1991) can be used to quantify uncertainty in inherent defect size in PVC-U pipes.

Probability Distribution of Inherent Defect Size in PVC-U Pipes

As suggested by Barker, Bowman and Bevis (1983), the Weibull probability distribution can model the variation in inherent defect size for plastic pipes reasonably well. A Weibull variable is one which has a "survivor function", $S(a)$, defined as:

$$S(a) = \exp\{-(a/\alpha)^{\eta}\} \tag{8.4}$$

where $S(a)$ is the probability that the variable a (inherent defect size) is greater than or equal to a particular value, α is the scale parameter of the Weibull distribution, and η is the shape parameter. Following Crowder et al. (1991), an empirical survivor function can be obtained as outlined below.

Suppose that all measured inherent defects have been put in ascending order $a_{(1)} < a_{(2)} \ldots < a_{(n)}$, where n is the number of measured defects. The empirical $S(a)$ is given by:

$$\hat{S}(a)_i = 1 - \frac{(i-1)}{n} \tag{8.5}$$

where i is the position of the ith defect size in the list arranged in ascending order (Crowder et al. 1991). Taking the logarithms of Equation 8.4 indicates that if the Weibull distribution is appropriate, a plot of $\ln\{-\ln S(a)\}$ versus $\ln(a)$ should be roughly linear.

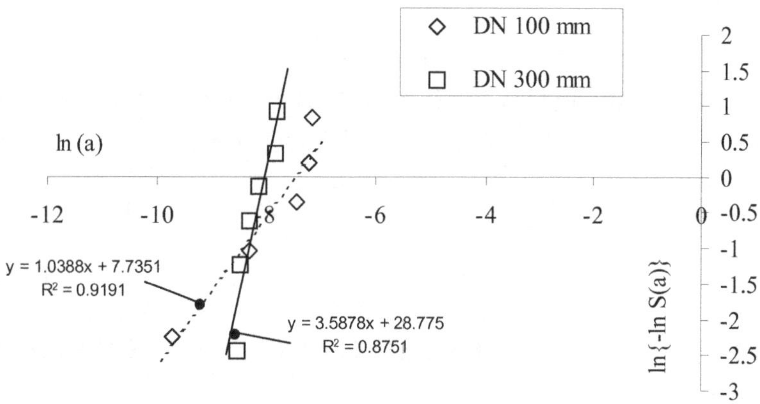

Figure 8.1. Weibull plot of inherent defect sizes for DN 100 mm and DN 300 mm Class 12 PVC-U Ca/Zn (AUS) pipes

The Weibull plot in Figure 8.1 shows inherent defect size data for both DN 100 mm and DN 300 mm, Class 12, PVC-U Ca/Zn (AUS) pipes. The data were obtained from SEM examinations of fracture surfaces from field failures in Australian PVC-U pipes. As shown, a straight line can be used to represent the data with regression coefficients of 0.92 and 0.88 for DN 100 mm and DN 300 mm pipe respectively. This suggests that the Weibull distribution represents the variation in defect size reasonably well in both cases. To further illustrate its applicability, Figure 8.2 shows a Weibull plot of combined data from DN 100 mm and DN 300 mm, Class 12, PVC-U Ca/Zn (AUS) pipes.

A linear fit to this combined data also provides correlation (r^2 coefficient = 0.95), indicating that the Weibull distribution represents variation in inherent defect size well. From Equation 8.7, the gradient and intercepts of the straight lines in Figures 8.1 allow the Weibull distribution parameters to be estimated. Values of η and α are given in Table 8.1.

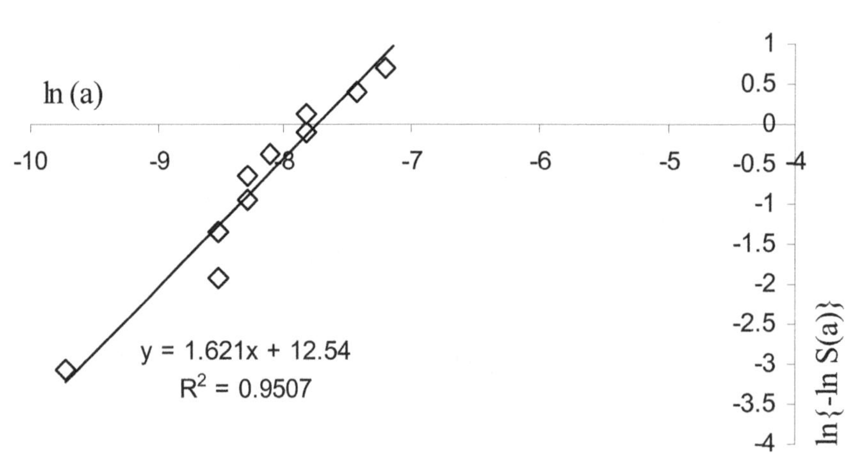

Figure 8.2 Weibull plot of inherent defect sizes: combined data from DN 100 mm and DN 300 mm, Class 12, PVC-U Ca/Zn (AUS) pipes

Table 8.1
Weibull parameters for inherent defect size in PVC-U pipe

Pipe type	Nominal diameter (mm)	Pressure class	Series	η (m)	α (m)
PVC-U Ca/Zn (AUS)	DN 100	12	II	1.0388	0.00058
	DN 300	12	II	3.5878	0.00033
	Combined data			1.6210	0.00044

SIMULATING PVC PIPE LIFETIMES USING MONTE CARLO SIMULATIONS

Having estimated the Weibull probability distribution for maximum *a*, Monte Carlo simulation can now be used to generate failures in a hypothetical PVC pipeline. The steps below outline the computational procedure used in this study:

1. Create a population of short segments (each 6-m long) of PVC pipes with known diameter and wall thickness.
2. Randomly assign an initial maximum defect size to each segment, based on the assumed Weibull probability distribution in Table 8.1. To provide a more accurate estimate of probable *a*, the Weibull distribution obtained from combined data is used.
3. Calculate remotely applied stresses for each pipe segment[2].
4. Calculate applied SIFs for each pipe segment[3].

---

[2] Internal pressure is calculated using Equation 7.4. Diametrical deflection is calculated using Equation 7.16 and the resulting tensile stress at the pipe inner surface is calculated from Equation 7.6. Values of residual stress at the pipe inner surface are simply assigned.

5. Calculate incubation periods for initial applied SIFs[4].

Time-marching loop

6. Increment time.
7. Calculate crack growth over time period[5].
8. Update applied SIFs[6].
9. Calculate the limit-state equations for brittle and ductile fracture[7].
10. Record pipe failures that occur, and replace failed segments with new segments.
11. Generate random inherent defect size in new pipe segments based on Weibull distribution as in Step 2.
12. Return to Step 5.

13. Write simulated lifetime data to output files after simulated time period ends.

As noted above, incubation times are calculated for each pipe segment. However, as demonstrated in Chapter 7, for pipes in service, applied SIFs are usually relatively low, which results in unrealistically high predicted incubation periods. This supports the earlier premise that all pipes contain a sharp notch, and that incubation times do not exist for pipes with sharp notches assessed at low SIFs. *For this reason, incubation time is omitted from the lifetime calculation, and lifetime is based on slow crack growth to failure only.*

The algorithm for Steps 1 through 12 has been implemented using the Python 2.3 programming language. A program listing, together with explanatory comments, is provided in Appendix D.

RESULTS FROM MONTE CARLO SIMULATIONS OF PVC PIPELINES

Using the Python program "PVCLifetime.py" (see Appendix D), lifetimes have been generated for a series of hypothetical pipelines. Failure in each pipeline segment has been predicted using the fracture mechanics model developed in Chapter 7. Both PVC-U Tin (USA) and PVC-U Ca/Zn (AUS) have been simulated under a range of operating conditions. In each case, fracture properties have been chosen based on the results obtained from the experimental study in Chapter 6.

---

[3] Applied SIFs for internal pressure, diametrical deflection, and residual stress are calculated from Equations 7.1, and 7.3–7.8.

[4] Incubation periods can be calculated from Equation 7.12. However, as outlined in Chapter 7, incubation periods are unrealistically high for many pipes in service, operating under relatively low applied SIFs. In these cases, incubation is recorded, but not used for lifetime calculation.

[5] Incremental crack growth over the time period, Δt, is calculated from Equation 7.14. Slow crack growth parameters are assigned for the particular pipe material under investigation.

[6] SIFs are updated using Equations 7.1, and 7.3–7.8.

[7] Limit states for brittle and ductile fracture are calculated from Equations 8.1 and 8.2 respectively.

Simulated Failure Data for PVC-U Tin (USA)

For PVC-U Tin (USA), the parameters listed in Table 8.2 were used as input to PVCLifetime.py. While these parameters were used in the examples in this chapter, it should be noted that they can be changed easily in the program PVCLifetime.py (Appendix D).

Table 8.2
Input parameters for Monte Carlo simulation of DN 4 inch, Class 150 psi, PVC-U Tin (USA) pipe

Parameter	Value
Number of segments in pipeline	20,000
Segment length (m)	6.0
Total simulated time (years)	110
Incremental time period (years)	1
Material short-term fracture toughness, K_{IC} (MPa.m$^{0.5}$)*	4.00[8]
Slow crack growth parameter, D	1.37×10^{-15}
Slow crack growth parameter, m	20.83
Inherent defect size – Weibull scale parameter α	1.621
Inherent defect size – Weibull shape parameter η	0.00043
Material short-term Young's modulus, E_0 (MPa)	1,775[9]
Visco-elastic parameter for reduction in Young's modulus, c	0.05
Material short-term yield strength, σ_0 (MPa)	29.25[10]
Visco-elastic parameter for reduction in yield strength, q	0.04
Outer radius, r_0 (m)	6.10×10^{-2}
Inner radius, r_i (m)	5.39×10^{-2}
Maximum internal pressure in each segment, p (MPa)	See simulated cases
Soil cover depth, H (m)	See simulated cases
Soil unit weight, γ (kN/m^3)	19.0[11]
Soil modulus, E_{soil} (MPa)	6.0[11]
Surface load, P_{surf} (kPa)	See simulated cases
Residual hoop stress acting at the pipe inner surface, σ_{IR} (MPa)	3.16

[8] From Table 6.3.
[9] Based on time in years; reduction in Young's modulus follows $E = E_0 t^{-c}$
[10] Based on time in years; reduction in yield strength follows $\sigma_\gamma = \sigma_0 t^{-q}$.
[11] Based on uncompacted sandy soil (AS/NZS 2566, 1998).

An average failure rate (per pipe length/year) can be defined for a particular pipe age as (Jarrett et al. 2001):

$$Average\ failure\ rate\ at\ age\ A = \frac{Number\ of\ observed\ failures\ at\ age\ A}{Exposed\ pipe\ length\ at\ age\ A} \tag{8.6}$$

For a particular pipe age, the exposed pipe length is that length of pipe that is the same age at the time of failure (Jarrett et al. 2001).

To allow simulated lifetime data to be converted into average failure rates, the PVCLifetime.py creates two output text files. The first file, "NewSegment.txt", lists the number of new pipe segments that were installed for each year of the Monte Carlo simulation. This allows the exposed pipe length to be calculated for a given pipe age and failure time. The second file, “FailHistory.txt”, lists the years when a particular pipe segment failed, which allows the number of failures to be recorded for a particular pipe age.

The curves in Figure 8.3 show the variation in simulated average failure rate with pipe age for PVC-U Tin (USA). Internal pressures of 0.8 MPa and 0.9 MPa were simulated, along with a cover depth of $H = 1$ m and a maximum surface load $P_{surf} = 20$ kPa.

The stochastic nature of inherent defect size is reflected in the simulated variation of average failure rate. However, it can be seen that in both cases, failure rate increases with pipe age. The systematic effect of an increase in internal pressure is also observed, with the maximum failure rates increasing from 1.2/100km/year for $p = 0.8$ MPa, to 2.8/100km/year for $p = 0.85$ MPa. An increase in pressure means that pipe segments exhibit higher crack growth rates, and in turn satisfy their failure criteria earlier.

As shown previously in Table 5.3, the field performance survey in Chapter 5 provided an estimated failure rate over a range of unknown pipe ages, diameters, and operating conditions for USA/Canadian water utilities. Monte Carlo simulations on DN 4 inch pipe ($p = 0.8$ MPa, $H = 1$ m, $P_{surf} = 20$ kPa) compare reasonably well to the upper and lower limits of actual recorded failure rate data (Figure 8.4).

For pipe aged up to approximately 60 years, the simulated average failure rates fall within the limits of 0.1 and 0.6 failures/100km/year for USA/Canadian water authority data. As pipe age increases, however, the simulated average failure rates increase above the upper limit.

While this agreement is encouraging, it should be noted that the operating conditions and pipe sizes associated with actual recorded failure data were not supplied by USA/Canadian water utilities. Values of internal pressure chosen for each Monte Carlo simulation were based on typical operating pressures for PVC-U pipes. However, in the absence of data describing the remaining operating loads, soil cover depth and surface loading were simply chosen by trial and error, until reasonable agreement was observed between the Monte Carlo simulations and actual failure rates reported by USA/Canadian water utilities. The applicability of the assumed burial depth $H = 1$ m and surface load $P_{surf} = 20$ kPa must be validated.

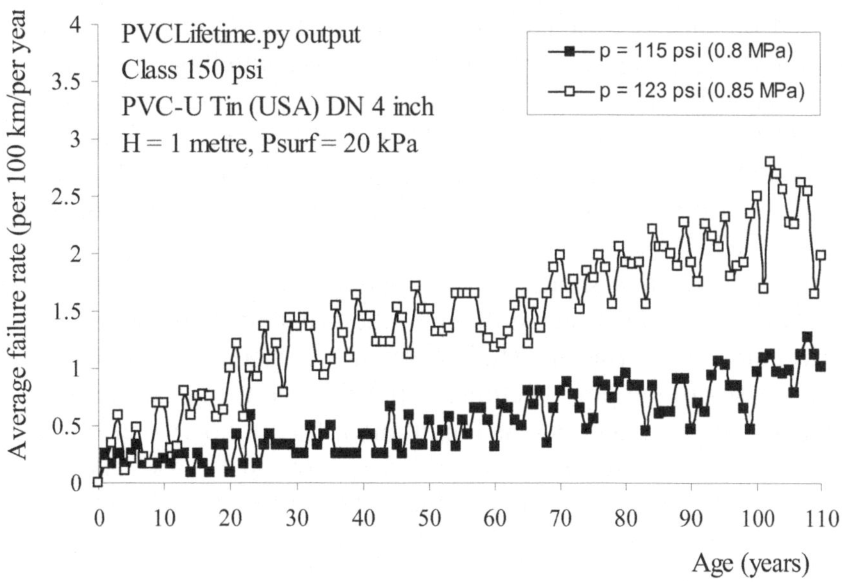

Figure 8.3 Simulated average failure rate versus age curve for DN 4 inch, Class 150 psi, PVC-U Tin (USA) pipe ($H = 1$ m, $P_{surf} = 20$ kPa): effect of internal pressure

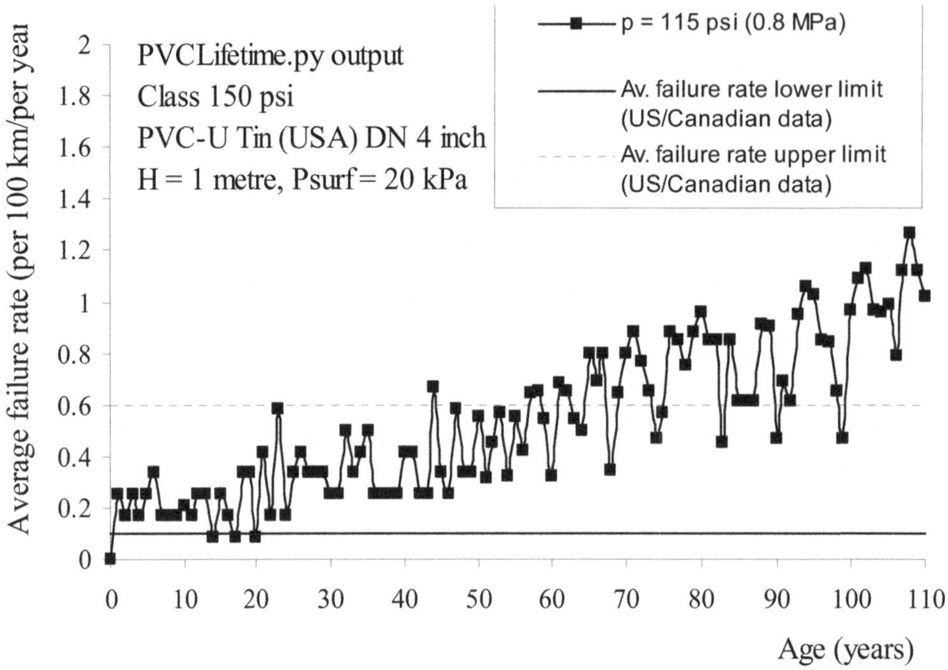

Figure 8.4 Simulated average failure rate versus age curve for DN 4 inch, Class 150 psi, PVC-U Tin (USA) pipe (H = 1 m, Psurf = 20 kPa): comparison with data from USA/Canadian water utilities

In addition to the soil weight and surface loads used in the Monte Carlo simulations, buried PVC pipes can be subjected to soil loads which vary with seasonal climate. For example, in a set of field tests on buried PVC pipes in Edmonton, Canada, Kuraoka, Rajani and Zhan (1996) measured circumferential and axial strains developed at the pipe outer surface over a one-year period. Based on measured axial and circumferential strains, an estimated mean normal pressure of 40 kPa was calculated. Depending on the surrounding soil support conditions, this may produce diametrical deflection and in-plane bending stresses. In another Canadian study, Rajani and Zhan (1996) used earth pressure cells to measure normal frost loads acting on buried PVC pipe. Over a 200-day period, maximum frost loads of 40 and 20 kPa were measured in clay and sand backfills respectively. Again, depending on the surrounding soil support, diametrical deflection and in-plane bending may occur.

An additional factor that should be considered when comparing simulated and observed failure rates for USA/Canadian water utilities is the breakdown of the reported failure modes. The Monte Carlo failure model in this study simulates in-plane loading only, and considers the action of tensile stresses in the pipe circumferential direction. In practice, this only corresponds to those failure modes that are reported as longitudinal fracture by water utilities. Referring back to Chapter 5, between 20% and 50% of recorded failures in USA/Canadian water utilities are attributed to longitudinal fracture. The remaining field failures are attributed to other causes, and are not captured by the Monte Carlo failure model.

Finally, it should also be noted that the Weibull distribution for maximum inherent defect size (used as input to the failure model) was actually obtained from field failures in *Australian* PVC pipes. As USA/Canadian field failure samples become available, the distribution of inherent defect sizes can be refined, and a more rigorous comparison of the failure model against the observed field performance can be made.

Simulated Failure Data for PVC-U Ca/Zn (AUS)

For comparison with PVC-U Tin (USA), a hypothetical pipeline comprised of segments of DN 100 mm, Class 1.6 MPa, PVC-U Ca/Zn (AUS) pipe was also simulated under the same operating conditions. With the exception of the parameters listed in Table 8.3, the same input parameters in Table 8.2 were used.

Figure 8.5 shows the effect of internal pressure on the simulated average failure rate in DN 100 mm, Class 1.6 Mpa, PVC-U Ca/Zn (AUS) pipe. Internal pressures of 0.8 and 0.9 MPa were simulated with $H = 0.4$ m and $P_{surf} = 15$ kPa in each case.

Table 8.3
Input parameters for Monte Carlo simulation of DN 100 mm, Class 1.2 MPa, PVC-U Ca/Zn(AUS) pipe

Parameter	Value
Material short-term fracture toughness, K_{IC} (MPa.m$^{0.5}$)	3.98[12]
Slow crack growth parameter, D	7.19×10^{-13}
Slow crack growth parameter, m	17.8
Outer radius, r_0 (m)	6.10×10^{-2}
Inner radius, r_i (m)	5.45×10^{-2}
Residual hoop stress acting at the pipe inner surface, σ_{IR} (MPa)	3.00

[12] From Table 6.3

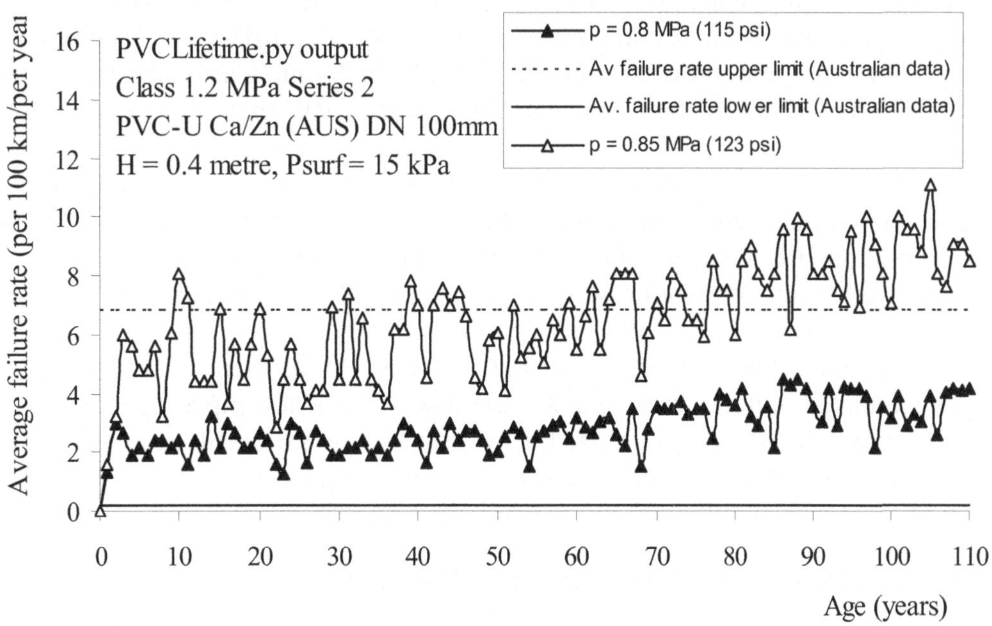

Figure 8.5 Simulated average failure rate versus age curve for DN 100 mm, Class 1.2 MPa, PVC-U Ca/Zn (AUS) pipe ($H = 0.4$ m, $P_{surf} = 15$ kPa): comparison with data from Australian water utilities

Similar to PVC-U Tin (USA), although random variations are observed, failure rates generally increase with pipe age in both cases. However, compared to the previous simulations, a relatively rapid initial increase in failure rate is observed, which then becomes more gradual with increasing pipe age. However, the effect of internal pressure is again clear. The maximum average failure rate increases from 3.2 failures/100km/year for p = 0.8 MPa, to 8 failures/100km/year for p = 0.85 MPa.

As shown, the model agrees reasonably well with the upper and lower limits from Australian water authority data (0.2 failures/100km/year, and 6.8 failures/100km/year; from Table 5.3). However, similar to the previous comparison for PVC-U Tin (USA), H and P_{surf} were chosen by trial and error to give reasonable agreement with recorded data at typical internal pressures. Data describing the installation and operating conditions of Australian PVC-U pipes are required to assess the applicability of the assumed H = 0.4 m and P_{surf} = 15 kPa. However, results from a study on buried PVC-U pipes in the UK suggest that these assumed service loads may be realistic (UKWIR 1997). Using buried load cells, the loads and temperatures in backfill adjacent to PVC-U pipes installed along a single-lane road were monitored. Similar to the Canadian study conducted by Kuraoka, Rajani and Zhan (1996), seasonal variations were observed, with the load increasing to a peak level in the middle of winter. In the UK study, soil backfill pressure increased by 15 kPa as ground temperatures decreased (Marshall 1997).

It should be noted that, similar to frost loading in the Canadian PVC pipes, PVC pipes in Australia may also be subjected to loads from expansive clay soils. It is well known that these soils undergo significant volume change upon moisture depletion, and (depending on surrounding soil support) may exert additional in-plane bending loads.

As with the previous comparison to USA/Canadian water authority data, the comparisons in Figure 8.5 include those other failure modes that are not captured by the Monte Carlo simulation model. However, a closer examination of the detailed failure and descriptions provided by Yarra Valley Water (YVW) allowed the variation in longitudinal fracture failure rate with pipe age to be isolated and examined. Since this data describe longitudinal fracture failures only, they provide a more rigorous benchmark for Monte Carlo simulations. As before, average rates of longitudinal fracture were calculated from YVW raw failure data using Equation 8.7.

The plot in Figure 8.6 compares average longitudinal fracture rate (failures/100km/year) obtained from YVW data and output from the Monte Carlo simulation model. The squares represent observations from YVW data. The vertical lines indicate the expected error associated with the number of events that comprise each failure rate. The error increases as the number of recorded events decreases.

As shown, for internal pressures of 0.8 and 0.85 MPa, there is reasonable agreement between the simulated failure rates and actual observations for pipes aged up to 20 years. According to YVW data, the longitudinal fractures in Figure 8.6 occurred at pressures between 0.4 and 1.2 MPa. While this is outside the lowest simulated pressure of 0.8 MPa, the agreement between the Monte Carlo simulations and actual failure rates is encouraging.

DEVELOPING FAILURE RATE CURVES FOR PVC-U PIPES

While the Monte Carlo simulation model can be used to investigate the influence of various key parameters on fracture failure rates in PVC-U pipes, simpler expressions relating failure rate to age are required for asset management. A useful technique to derive failure rate versus age curves is to determine the probability distribution that best models the variation in simulated lifetimes.

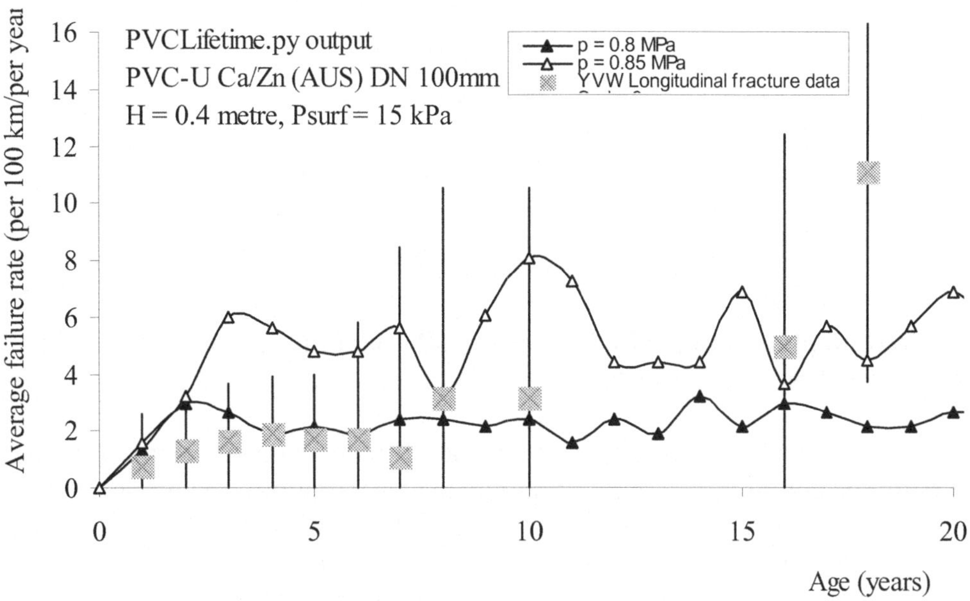

Figure 8.6 Simulated average failure rate versus age curve for DN 100 mm, PVC-U Ca/Zn (AUS) pipe (H = 0.4 m, P_{surf} = 15 kPa): comparison with data from Yarra Valley Water longitudinal fracture failure data

Figures 8.7 and 8.8 show Weibull plots for PVC-U Tin (USA) and PVC-U Ca/Zn (AUS) pipe at internal pressures of 0.8, 0.85 and 0.9 MPa. In each case, lifetimes were simulated for 100 pipe segments. $S(t)$ is the empirical survivor function for lifetime, as defined previously in Equation 8.3. As shown, lifetimes from Monte Carlo simulations are reasonably well represented by a straight line, illustrating the applicability of the Weibull distribution.

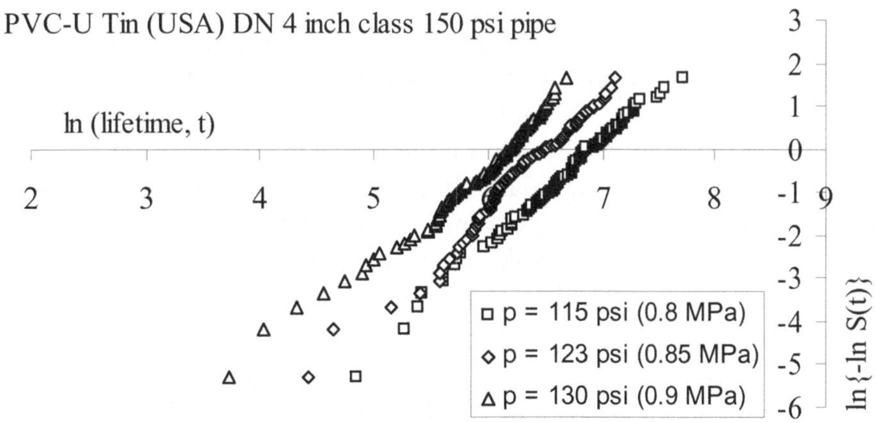

Figure 8.7 Weibull plots: sampled lifetimes for DN 4 inch, Class 150 psi, PVC-U Tin (USA) pipe (H = 1 m, P_{surf} = 20 kPa)

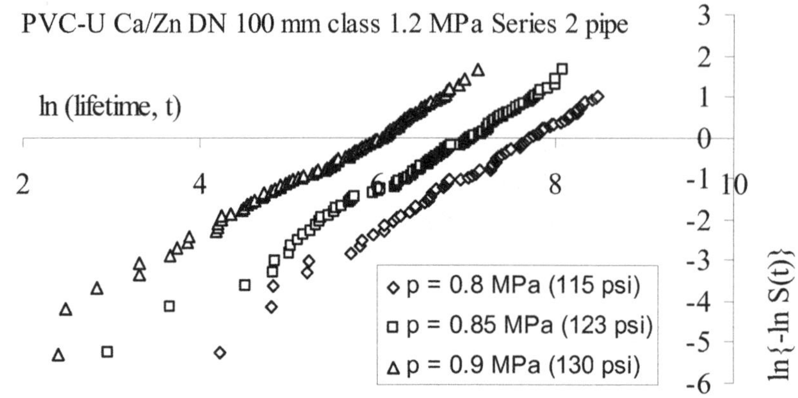

Figure 8.8 Weibull plots: sampled lifetimes for DN 100 mm, Class 1.2 MPa, PVC-U Ca/Zn (AUS) pipe ($H = 0.4$ m, $P_{surf} = 15$ kPa)

By making the simplifying assumption that all pipe segments along a pipeline have the same probability of failure (i.e., assuming homogeneous variation), the number of failures along the pipelines can be described by the binomial probability function (Crowder et al. 1991). This allows the expected failure rate (per unit pipe length/time period) to be estimated. If each individual pipe segment sustains one failure, which is independent of other failures, the expected number of failures in the pipeline is:

$$E(N) = nh(t) \tag{8.7}$$

where n is the number of segments in the pipeline, and $h(t)$ is the hazard/failure rate function (Crowder et al. 1991). In practice, $h(t)$ is a measure of the probability of imminent failure at time t, given that failure has not yet occurred (Crowder et al. 1991). For the Weibull distribution, $h(t)$ is defined in terms of the scale parameter α, and the shape parameter η as:

$$h(t) = \eta\alpha^{-\eta}t^{\eta-1} \tag{8.8}$$

Since each pipe segment in the Monte Carlo simulation is randomly assigned an initial defect size, assuming the same probability of failure (as required by the binomial probability function) is unrealistic. However, as shown in Figures 8.9 and 8.10, the resulting failure rate versus age curves show good agreement with failure rates from the Monte Carlo simulation.

In Figure 8.9, raw failure rate data (from the Monte Carlo simulation) is compared to expected failure rates from the corresponding Weibull hazard function for DN 4 inch, Class 150 psi, PVC-U Tin (USA) pipe. In both cases, the expected number of failures is calculated from Equations 8.6 and 8.7. For a 100-km length of pipeline, the number of segments n is 1.67×10^4 (each segment 6-m long).

As shown in Figure 8.10, good agreement is also obtained from DN 100 mm, Class 1.6 MPa, PVC-U Ca/Zn (AUS) pipe. The Weibull hazard function also captures the relatively sharp initial increase in failure rate, and the subsequent gradual increase with pipe age.

Table 8.4 summarizes the Weibull lifetime distribution parameters for a range of different operating pressures. While different pressures were simulated, the cover depths and maximum surface loads were $H = 1$ m and $P_{surf} = 20$ kPa for PVC-U Tin (USA), and $H = 0.4$ m and $P_{surf} = 15$ kPa for PVC-U Ca/Zn (AUS)

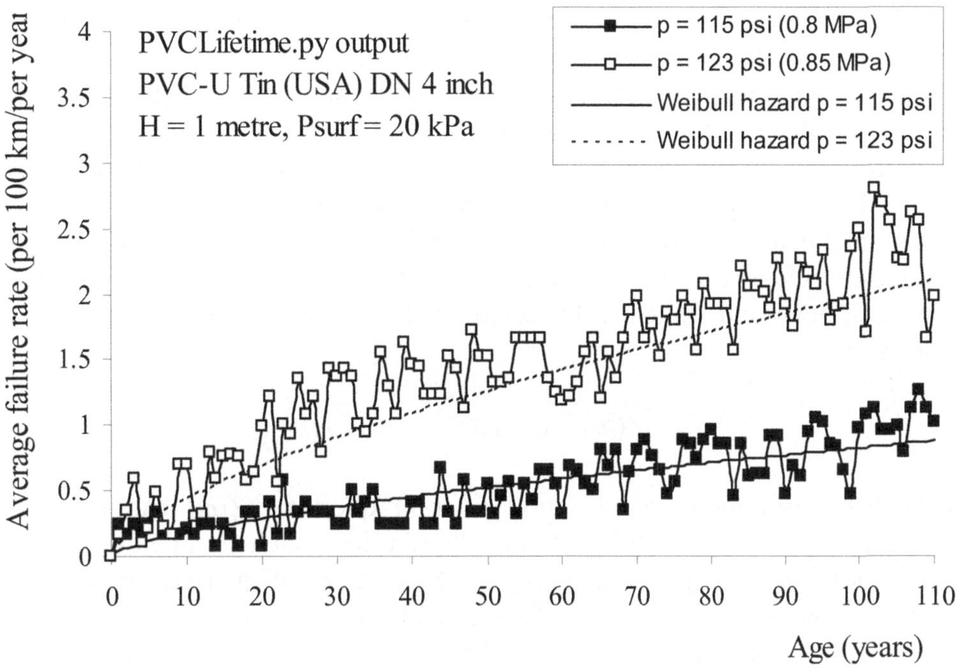

Figure 8.9 Simulated average failure rate versus age curve for DN 4 inch, Class 150 psi, PVC-U Tin (USA) pipe ($H = 1$ m, $P_{surf} = 20$ kPa): comparison of raw data from Monte Carlo simulation and Weibull hazard function

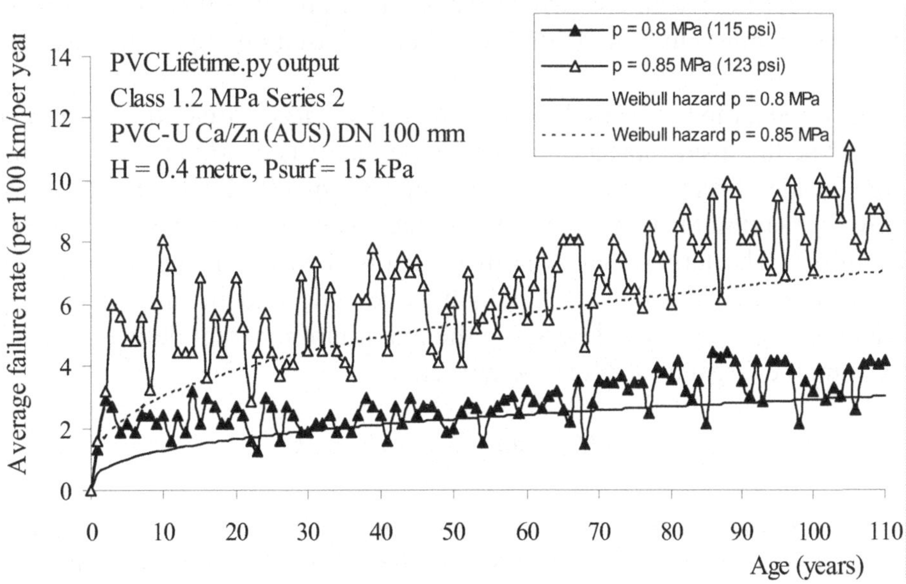

Figure 8.10 Simulated average failure rate versus age curve for DN 100 mm, Class 1.2 MPa, PVC-U Ca/Zn (AUS) pipe ($H = 0.4$ m, $P_{surf} = 15$ kPa): comparison of raw data from Monte Carlo simulation and Weibull hazard function

THE EFFECT OF PIPE SIZE AND OPERATING PRESSURE ON AVERAGE FAILURE RATE

The effect of pipe size on failure rate vs. age curves can be analysed using the size dependence of the Weibull distribution for inherent defect size.

Having fitted the inherent defect size results to an appropriate distribution, we must now extrapolate these results to different pipeline sizes. It might not be immediately obvious that the probability distribution for defect size will change when viewing a larger area. The reason for this change is that extreme values (e.g. minimum or maximum values) of defect size are modelled. As an illustrative example, assume for instance that the lowest price on a certain product is required. Queries can be made to different shops to get price quotes, which will be distributed with a mean and standard deviation. If the quotes obtained are x_1, x_2, x_3.. etc., the interesting quantity is

$$Z(m) = \min(X_i) \quad i = 1 \ldots m,$$

which is the minimum quote found after m queries. $Z(m)$ can be assumed to be distributed according to some extreme value distribution (such as Weibull). Intuitively, the probability of finding a good price increases as the number of queries made. In practice, this means that the distribution for $Z(m)$ also changes with the number of queries, m, that are made. The problem therefore is to quantify how the distribution for $Z(m)$ changes with m.

In a similar sense, there are a number of inherent defects in each segment of a pipeline. In sampling, we choose a number of sample areas and retrieve a set of sample data from each, giving the maximum pit depth. We then estimate an Extreme Value distribution for the maximum defect size within the sample area, which has an assumed number of defects equal to m (this is the same as making m queries in the pricing example above). If we want to find the distribution for a larger pipe segment (which has a larger area and therefore a greater number of pits) it follows that we must find the related distribution based on the quota between the relevant areas.

Consider for example a sample area, A_s and a target area for extrapolation, A_t. In practice, A_s would be the inside surface area of a pipe that was examined at each sample site, and A_t could be the inside surface area of a larger pipe segment. Each of the sampling areas is assumed to have a maximum defect size, X_i (a stochastic variable) which is described by the Weibull distribution. The target area A_t is then idealised as consisting of individual segments, (each with a surface area equal to A_s), which have independent and identically distributed stochastic variables, (X_i; $i = 1 \ldots m$), describing the maximum inherent defect size. The maximum defect size Y within the target area A_t can be written as

$$Y(k) = \max[X_i] \quad i = 1 \ldots k$$

where k is the total number of segments within A_t ($k = A_t/A_s$) and X_i is the maximum pit depth within a certain segment. According to the theory for Weibull distributions (Blom 1989, Weibull 1948), the probability distribution function for the maximum defect size within the target area can then be found using the transformations in equation (8.9) and (8.10).

$$\alpha_Y = \alpha_X k^{\left(\frac{1}{\eta_X}\right)} \tag{8.9}$$

$$\eta_Y = \eta_X \tag{8.10}$$

The Weibull distribution parameters α_X and η_X in eqs. (8.9) and (8.10) apply to A_s and the parameters η_Y and α_Y apply to A_t. Hence, the probability distribution function for inherent defect size in the sample area can be extrapolated to a larger target area based on their area ratio. It should be noted that the main underlying assumption in this transformation is that the number of pits in a particular segment, is proportional to the area of that segment. In the absence of sufficient defect size data from pipe field failures, the validity of this assumption is uncertain.

However, assuming (as a first approximation) that defect size distributions can be scaled according to the ratio between sample and target areas, Weibull scale and shape parameters can be estimated as in Table 8.4.

Table 8.4
Weibull scale and shape parameters for inherent defect size distributions in different nominal diameters of Class 150 psi PVC-U (Tin) (USA) pipes

Class 150 psi (1.03 MPa) PVC-U (Tin) (USA)			
Nominal diameter (inches)(mm)	Ratio of target and sample areas $k = A_t/A_s$	Weibull distribution for defect size (in metres)	
4.0 (101.6)	1.00	Scale α	0.00044
		Shape η	1.621
6.0 (152.4)	1.55	Scale α	0.00057
		Shape η	1.621
8.0 (203.2)	2.04	Scale α	0.00068
		Shape η	1.621
10.0 (254.0)	2.49	Scale α	0.00077
		Shape η	1.621
12.0 (304.8)	2.96	Scale α	0.00086
		Shape η	1.621

The parameters in Table 8.5 show the effect of pipe size on the defect distribution in Australian Class 1.2 MPa PVC pipes.

Table 8.5
Weibull scale and shape parameters for inherent defect size in different nominal diameters of Class 1.2 MPa Series 2 PVC-U (Ca/Zn) (AUS) pipes

Class 1.2 MPa (173.9 psi) Series 2 PVC-U (Ca/Zn) (AUS)			
Nominal diameter (mm)(inches)	Ratio of target and sample areas[8] $k = A_t/A_s$	Weibull distribution for defect size (in metres)	
100 (3.94)	1	Scale α	0.00044
		Shape η	1.621
150 (5.91)	1.55	Scale α	0.00058
		Shape η	1.621
200 (7.87)	2.04	Scale α	0.00069
		Shape η	1.621
225 (8.86)	2.49	Scale α	0.00074
		Shape η	1.621
250 (9.84)	2.96	Scale α	0.00079
		Shape η	1.621
300 (11.81)	3.03	Scale α	0.00089
		Shape η	1.621

Having extrapolated Weibull parameters, Monte Carlo simulations can be done for different pipe sizes, at different pressures. For example, the plots in Figures 8.11 and 8.12 show failure rate vs. age curves for DN 8 inch and DN 12 inch Class 150 psi USA PVC pipe.

In comparison to DN 4 inch pipe, simulated failure rates increase as pipe size increases. As discussed above, extrapolation of the defect size distribution to larger pipe sizes results in a larger defect size assigned to pipe segments in the Monte Carlo simulation. This is reflected in the increased Weibull scale parameters in Tables 8.4 and 8.5. Consequently, for a given operating pressure, segment lifetime is reduced and the expected failure rate increases.

Having defined the basis for extrapolating defect size distributions to larger pipe sizes, generic equations can be defined for failure rate that are dependent on pipe size and operating pressure. To achieve this, Monte Carlo simulations are run for different pipe size and operating pressure combinations, and failure rate vs. age curves are generated. Equations (8.6) and (8.7) are then parameterized in terms of pipe size and operating pressure, using a least squares regression. The failure rate vs. age curves from Monte Carlo simulations of different size and

[8] Based on the inside surface area of the pipe

pressure combinations are given in Appendix E. It should be noted that the Weibull defect size distribution was extrapolated over a restricted range of pipe sizes only:

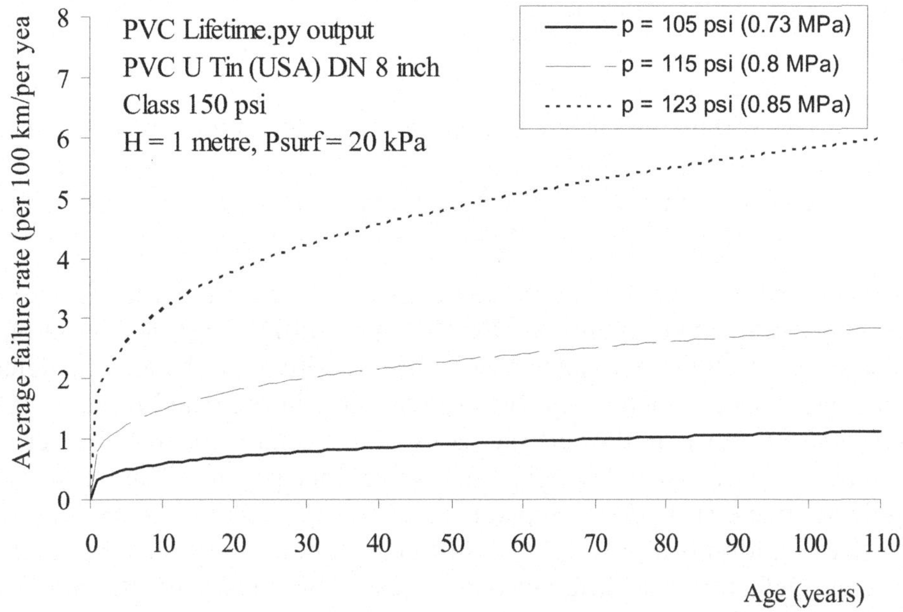

Figure 8.11 Simulated average failure rate versus age curve for DN 8 inch, Class 150 psi, PVC-U Tin (USA) pipe (*H* = 1 m, P_{surf} = 20 kPa)

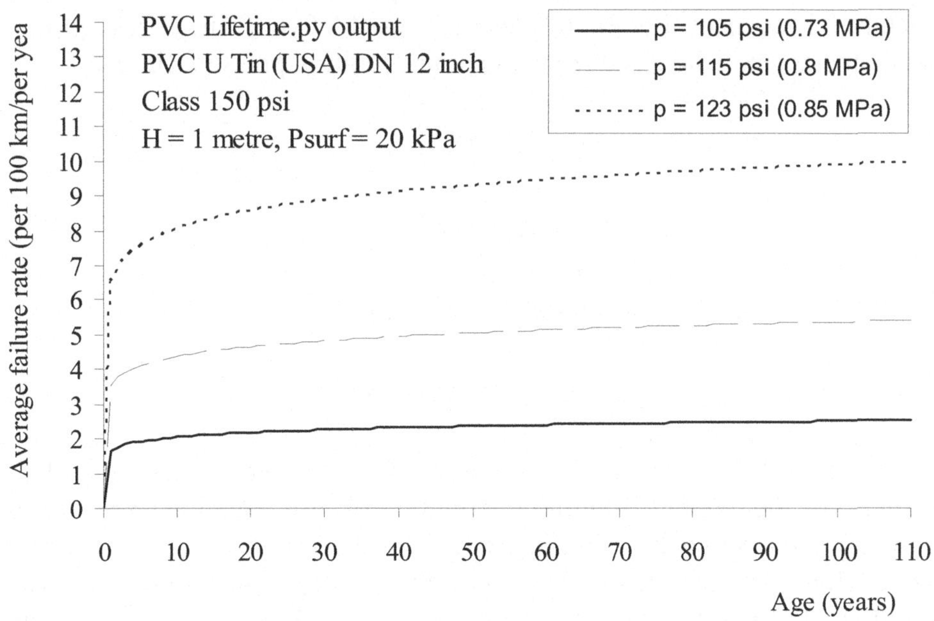

Figure 8.12 Simulated average failure rate versus age curve for DN 12 inch, Class 150 psi, PVC-U Tin (USA) pipe (*H* = 1 m, P_{surf} = 20 kPa)

Table 8.6
Size ranges for extrapolation of Weibull maximum defect size distribution

	Pipe type	Size range
USA	Class 150 psi PVC-U (Tin) (USA)	DN 4 inch to DN 12 inch
Australia	Class 1.2 MPa PVC-U (Ca/Zn) (AUS)	DN 100 mm to DN 300 mm

The reason for this limited extrapolation is that (as shown in Figure 8.1) measured defect sizes were obtained from field failures in DN 100 mm and DN 300 mm pipes only. Lifetime predictions based on fracture mechanics theory are sensitive to changes in defect size, with relatively small increases producing significant reductions in predicted lifetime. Consequently, for larger pipe sizes (with extrapolated increases in maximum inherent defect size) forecast failure rates may be unrealistically high. To solve this problem (and test the validity of extrapolating defect size distributions), separate databases of defect sizes should be generated for different pipe sizes. Individual probability distribution functions can then be determined for the maximum inherent defect size in different pipe types. This illustrates the importance of water utilities storing samples from field failures in PVC pipes and the examination of fracture surfaces to identify inherent defect size.

However, as a first approximation, the extrapolation of defect size distribution obtained in this study is assumed to be valid over the pipe size ranges shown in Table 8.6. Equations (8.11) to (8.14) describe the effect of pipe size and operating pressure on the Weibull scale and shape parameters for segment lifetime of class 150 psi PVC-U (Tin) (USA) pipe. (DN) is the pipe nominal diameter (in inches) and (p) is the pipe operating pressure (in psi)

$$\eta = 2.8(DN)^{-0.38} \tag{8.11}$$

$$\log_{10}\alpha = m(p) + c \tag{8.12}$$

$$m = 0.0003(DN) - 0.034 \tag{8.13}$$

$$c = -0.054(DN) + 7.58 \tag{8.14}$$

Equations 8.11–8.14 estimate the pressure and pipe size dependence of the scale and shape parameters for the Weibull lifetime distribution. These equations can now be combined with Equations 8.7 and 8.8 to generate failure rate vs. age curves for different operating pressures and pipe sizes. However, as stated in Table 8.6, the resultant curves obtained using equations (8.7), (8.8) and (8.11) to (8.12) will only apply to class 150 psi pipes with nominal diameters between DN 4 inches and DN 12 inches.

A similar set of parameterised equations for Weibull lifetime distribution parameters can be obtained for Class 1.2 MPa, series 2 PVC-U (Ca/Zn) (AUS) pipes.

$$\eta = 1.96(DN/25.4)^{-0.27} \tag{8.15}$$

$$\log_{10}\alpha = m(144.9p) + c \tag{8.16}$$

$$m = -0.034 \tag{8.17}$$

$$c = -0.062(DN/25.4) + 7.55 \tag{8.18}$$

In this case, the pipe nominal diameter (DN) is in mm and operating pressure (p) is in MPa. As with USA PVC, equations (8.15) to (8.18) can be combined with equation (8.7) and (8.8) to estimate expected average failure rate vs. age curves. However, as stated in Table 8.6, these curves will only apply to class 1.2 MPa Series 2 pipes with nominal diameters between DN 100 mm and DN 300 mm. Plots of estimated failure rate vs. age for Class 1.2 MPa Series 2 PVC-U (Ca/Zn) (AUS) pipes are given in Appendix E.

COMPARING PREDICTIONS WITH ACTUAL FAILURE RATES IN PVC-U PIPES FROM THE UKWIR DATABASE

From fracture toughness data published by Marshall (1982), similar values are obtained for UK PVC-U pipes and PVC-U Ca/Zn (AUS). Since the development of PVC-U pipes in Australia closely followed developments in the UK (Burn and Richards, 1995), it is not surprising that both these pipe materials exhibit similar fracture toughness characteristics. With this in mind, the UKWIR national database of mains failures provides a good benchmark to compare predicted failure rates form the physical probabilistic failure model for PVC-U pipes.

As before, an average failure rate (per pipe length/year) can be defined for a particular pipe age as (Jarrett et al. 2001):

$$\textit{Average failure rate at age A} = \frac{\textit{Number of observed failures at age A}}{\textit{Exposed pipe length at age A}} \tag{8.19}$$

For a particular pipe age, the exposed pipe length is that length of pipe that is the same age at the time of failure (Jarrett et al. 2001).

As described in Chapter 5, average failure rates (failures/100km/per year) for different pipe ages can be extracted form the UKWIR database using the Python 2.3 program in Appendix C. To ensure a fair comparison, only those failures that occurred by longitudinal fracture are extracted from the database. As before, 95% confidence limits associated with each estimate of average failure rate is approximately given by (Jarrett et al. 2001)

$$\text{Upper Bound} = \text{Av. Failure Rate} \times \left[\exp\left(\frac{2}{\sqrt{n}}\right)\right] \tag{8.20}$$

$$\text{Lower Bound} = \text{Av. Failure Rate} \div \left[\exp\left(\frac{2}{\sqrt{n}}\right)\right] \tag{8.21}$$

where *n* is the number of observed failures that comprise the estimated average failure rates. Since PVC-U pipes were produced with relatively low levels of gelation (and hence fracture toughness) before 1965 (Alferink, Janson and Holloway 1997), comparisons are restricted to pipes in the UKWIR database that are less than 40 years old.

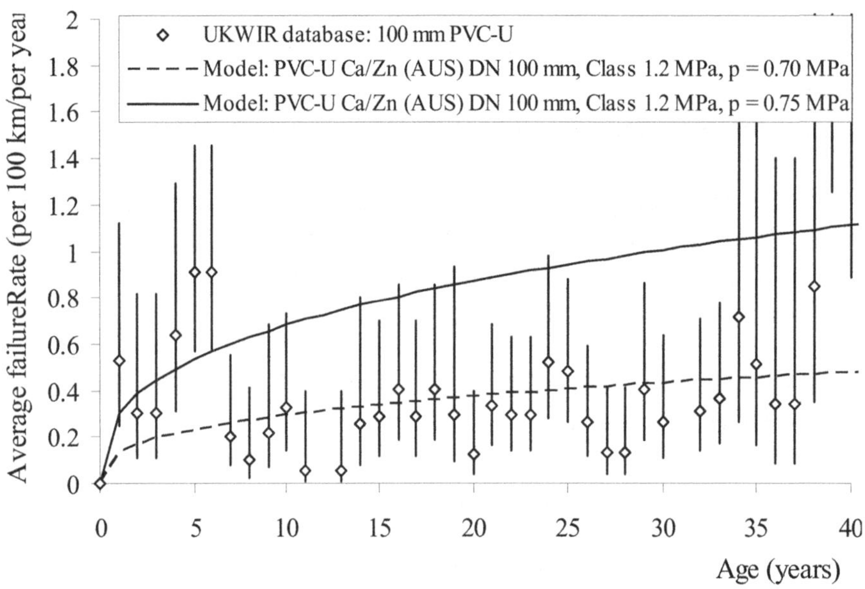

Figure 8.13 Comparison between predicted failure rates in DN 100 mm, Class 1.2 MPa PVC-U Ca/Zn (AUS) pipe and observed failure rates in DN 100 mm PVC-U pipes in UKWIR database. Predicted failure rates from eqs. (8.6), (8.7) and (8.14) to (8.17)

For comparison with actual data, curves of predicted average failure rate vs. age at operating pressures of 0.7MPa (= 101psi) and 0.75 MPa (109 psi) are plotted in Figure 8.13.

While the shape of the UKWIR failure rate curve may not be captured fully, both predicted curves show reasonable agreement within the 95% confidence limits. It should also be noted that actual operating pressures are not supplied in the UKWIR database and the values of 0.7 MPa (101 psi) and 0.75 MPa (109 psi) used to predict failure rates are merely based on realistic operating pressures in Australian water utilities. Failure data with actual operating pressures is required for a more rigorous comparison.

Comparisons between actual and predicted failure rates for larger diameter PVC-U pipes in the UKWIR database are shown in Figures 8.14 to 8.17.

Again, the predicted curves at pressures of 0.7 MPa (101 psi) and 0.75 MPa (109 psi) do not capture the actual shape of the UKWIR failure rate curve. However, they fall within the 95% confidence limits suggesting that the extrapolation of Weibull distribution parameters for maximum inherent defect size (and the resulting equations 8.15 to 8.18 accounting for pipe size effects on failure rate) are reasonable. However, examination of fracture surfaces from field failures in different sized pipes is required for validation.

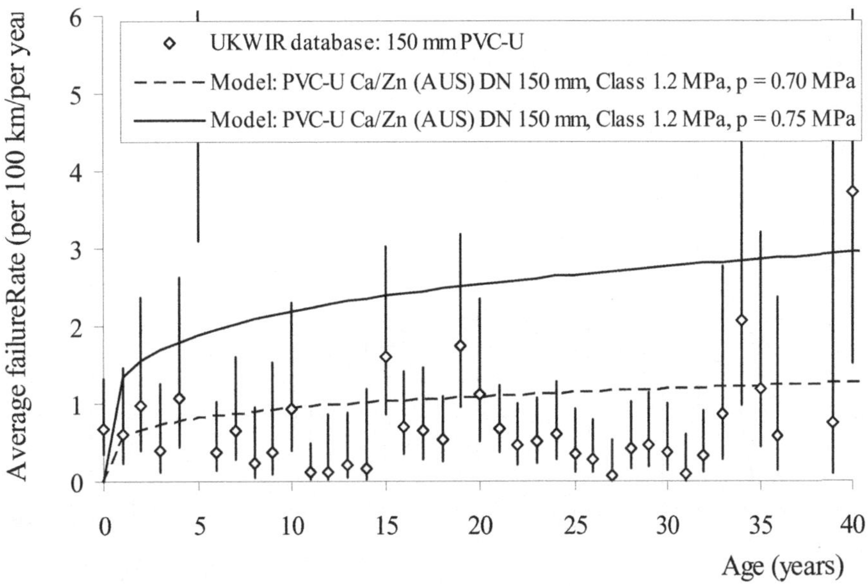

Figure 8.14 Comparison between predicted failure rates in DN 150 mm, Class 1.2 MPa PVC-U Ca/Zn (AUS) pipe and observed failure rates in DN 150 mm PVC-U pipes in UKWIR database. Predicted failure rates from eqs. (8.6), (8.7) and (8.14) to (8.17)

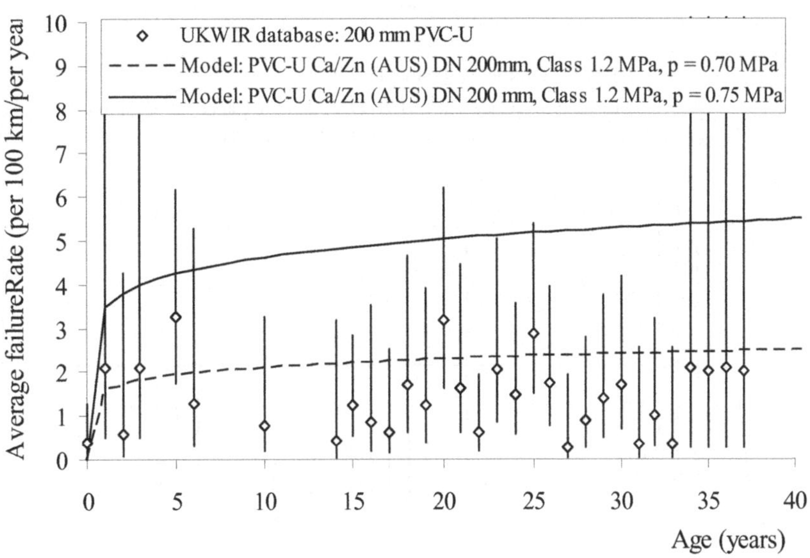

Figure 8.15 Comparison between predicted failure rates in DN 200 mm, Class 1.2 MPa PVC-U Ca/Zn (AUS) pipe and observed failure rates in DN 200 mm PVC-U pipes in UKWIR database. Predicted failure rates from eqs. (8.6), (8.7) and (8.14) to (8.17)

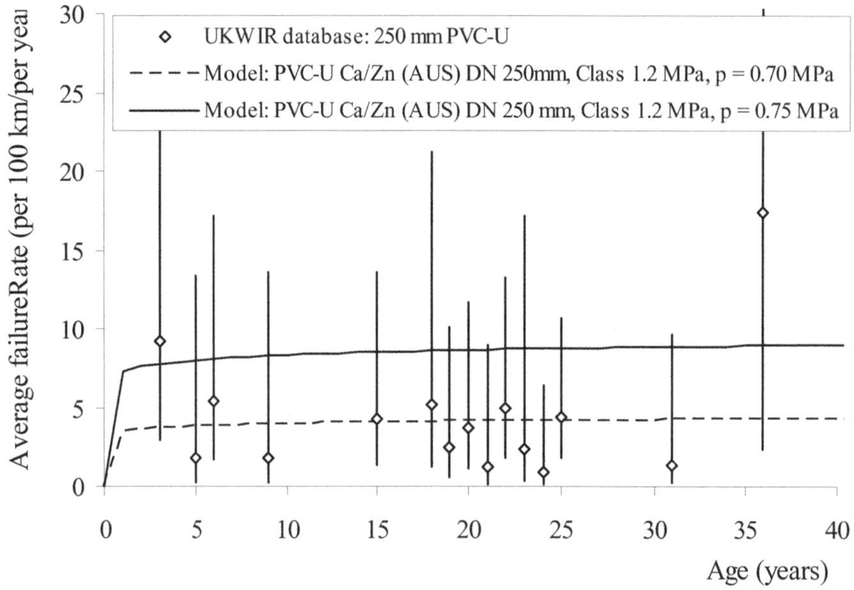

Figure 8.16 Comparison between predicted failure rates in DN 250 mm, Class 1.2 MPa PVC-U Ca/Zn (AUS) pipe and observed failure rates in DN 250 mm PVC-U pipes in UKWIR database. Predicted failure rates from eqs. (8.6), (8.7) and (8.14) to (8.17)

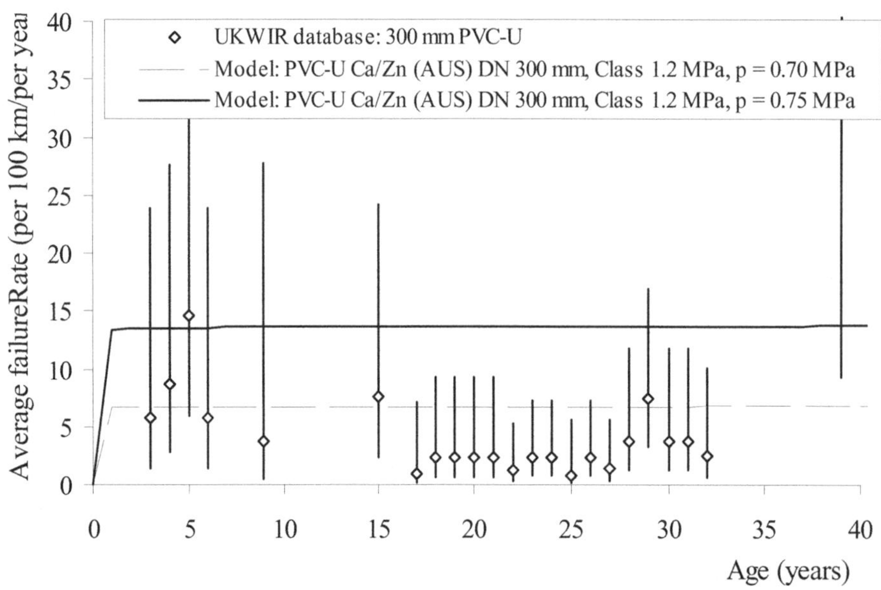

Figure 8.17 Comparison between predicted failure rates in DN 300 mm, Class 1.2 MPa PVC-U Ca/Zn (AUS) pipe and observed failure rates in DN 300 mm PVC-U pipes in UKWIR database. Predicted failure rates from eqs. (8.6), (8.7) and (8.14) to (8.17)

CONCLUSIONS & RECOMMENDATIONS

To capture uncertainty in the pipe fracture process, a Monte Carlo simulation model has been developed for PVC pipelines. Based on the fracture mechanics failure model developed in Chapter 7, the Monte Carlo simulation generates a hypothetical pipeline, comprised of a series of independent pipe segments. The maximum inherent defect size in each pipe segment is treated as a stochastic variable, which (based on examination of fracture surfaces from field failures) is represented by the Weibull probability distribution function.

For each segment in the hypothetical pipeline, the Monte Carlo simulation randomly assigns a maximum inherent defect size, based on the Weibull distribution. A time-marching loop is then performed in which incremental crack growth is simulated in each pipe segment. Failure is assumed to occur by brittle fracture when the applied SIF exceeds the material fracture toughness. An alternative criterion for ductile fracture is satisfied when the applied net section stress exceeds the material yield strength. Upon failure, a new pipe segment is installed and an inherent defect assigned. Output from the Monte Carlo simulation provides the number of new pipe segments installed for each year of the simulation, and the failure years for each pipe segment. Combining these data allows the average failure rate (failures/length/year) to be estimated. Making the simplifying assumption that the number of failures along the pipeline followed a binomial process, allowed the expected number of failures to be described by the Weibull hazard (or failure rate) function.

To assess the effect of pipe size on average failure rate, the Weibull distribution parameters for inherent defect size were extrapolated based on increases in the pipe inside surface area. Monte Carlo simulations were then run for different pipe size/pressure scenarios and a simple set of 'empirical' equations were proposed, which expressed average failure rate in terms of operating pressure and pipe size. However, it should be noted that these equations should be restricted to the following pipe sizes:

- Class 150 psi PVC-U (Tin) (USA) pipes, DN 4 inch to DN 12 inch
- Class 1.2 MPa, Series 2 PVC-U (Ca/Zn) (AUS) pipes, DN 100 mm to DN 300 mm

A simple set of "look up" graphs were also compiled in Appendix E of this report. For comparison with actual failure rates recorded by water utilities, average failure rate vs. age curves were extracted form the UKWIR database. To ensure a fair comparison, only longitudinal fracture failures in pipes less than 40 years old were included. It is likely that older pipes will have been poorly processed and exhibited lower fracture toughness levels. Since fracture properties of UK and Australian PVC-U are reportedly similar (Marshall, 1982), actual failure rates form the UKWIR database were compared to predicted failure rates for PVC-U Ca/Zn (AUS) Class 1.2 MPa pipes. While the shape of the UKWIR failure rate curve was not captured fully, predicted curves showed reasonable agreement within the 95% confidence limits for actual failure rates in diameter between 100 and 300 mm.

Compared to other probabilistic modeling techniques, Monte Carlo simulation allows complex limit-state failure models to be used without the need for simplification. In contrast, techniques such as FOSM methods require linear approximations to failure models, and the assumption that stochastic variables are normally distributed. Consequently, the realism of the failure process can be compromised. However, while the Monte Carlo simulation model could be useful, several areas are identified for future development.

Inherent Defect Size

It should be noted that only those fracture failures that occurred in Australian PVC-U pipe were examined in this study. Consequently, the applicability of the Weibull distribution for inherent defect size to USA/Canadian PVC pipes is not yet validated. To this end, pipe samples from USA/Canadian field failures should be exhumed and the fracture surfaces examined. While an initial examination under optical microscope will reveal the origin of crack initiation, SEM should be used to quantify the inherent defect size.

It should also be noted that defect size data in this study were obtained from a range of pipe sizes. If these inherent defects are created during the pipe manufacturing process (cavities for example), their size distribution may also depend on pipe diameter and wall thickness. To this end, a continued program of field failure examination is required to determine the influence of pipe size on inherent defect size.

Applied Stress Intensity Factors for Different Defect Shapes and Locations

The failure model developed in this study describes SIFs in terms of radial defect size only. To provide a worst-case scenario, defects have also been assumed to be located at the pipe inner surface, where circumferential stresses from internal pressure, diametrical deflection, and residual stress are highest. In practice, a range of defect shapes and pipe wall locations will exist, and SIF solutions are available (Rooke and Cartwright 1976). However, while these improvements to the physical model could be readily incorporated into the Monte Carlo simulation, there are currently insufficient data to describe probability distributions for defect size and shape. A continued program of field failure examination is required to solve this limitation.

Loading Conditions

As outlined in Chapters 7 and 8, the fracture mechanics failure model requires the applied circumferential stresses due to internal pressure, external loading, and residual stress as input. While internal pressure and residual stress may be well defined, external loading is idealized in the model as a soil weight and a surface load acting above a buried pipe. A more realistic model should account for external loads generated from frost (in cold climates) and expansive clays (due to moisture depletion). For example, Rajani and Zhan (1996) propose an analytical model to estimate the magnitude of frost loads in cold climates. The model predicts the change in earth pressure as a frost front propagates, and shows good agreement with measured values. By considering the combined stiffness of a buried pipe and surrounding soil, it may be possible to calculate the magnitude of normal frost loads on a buried pipe, and bending stresses generated in the pipe wall. In a UK study, earth pressures in soil backfill adjacent to buried PVC-U pipes were recorded to increase by 15 kPa during winter. However, with minimum ground temperatures recorded at 6°C (43F) (Marshall 1997), frost loading did not contribute to this load change. An alternative mechanism for this load increase may be the volumetric increase of expansive clay soils on moisture absorption. Again, depending on the surrounding soil support, this could create bending stresses in the wall of a buried pipe. A similar approach to that adopted for predicting frost loads may also be possible to model the action of expansive soils on buried pipes.

Since the Monte Carlo simulation method allows for a great deal of complexity in the limit failure models, more accurate descriptions of external loading under frost or expansive clays can be readily incorporated, provided analytical models can be developed.

Crack Incubation Times for Pipes in Service

As outlined in Chapter 7, initial applied SIFs for pipes in service are relatively low, which gives unrealistically high incubation periods predicted from fracture mechanics theory. While predicted incubation times are reasonable for higher applied SIFs, an alternative approach is required to address pipes in service. Following previous work conducted at CSIRO, an experimental program to observe incubation periods for a range of defect types and applied SIFs should be conducted. This will validate the current evidence suggesting that incubation periods do not exist for pipes operating under low applied SIFs in service.

Different Failure Modes

Although reasonable agreement was possible between the Monte Carlo simulation model and recorded failure rate data, it should be noted that the simulation accounts only for those failures that occur by longitudinal fracture. Consequently, these longitudinal fracture failures should be isolated from the total recorded failure rate. Since failure rates in USA/Canadian water utilities are low, this may only be practical for Australian utilities. Although initial comparisons with longitudinal fracture data from YVW are encouraging, the applicability of external loads used in the Monte Carlo simulation should be validated.

Generic Failure Rate Versus Age Curves

The current study has demonstrated that the Weibull distribution function is well suited to describing the variation in lifetimes generated from the Monte Carlo simulation. The graphs in Figures 8.11 and 8.12 illustrate the pressure dependence of the Weibull scale and shape parameters that describe the lifetime distribution. However, it is reiterated that in the current study, *particular pipe sizes and external loading conditions were chosen*. For other situations, varying these conditions and repeatedly running the Monte Carlo simulation model presented in this study can explore the dependence of Weibull lifetime distribution parameters on pipe size, internal pressure, and external loading conditions. By generating graphs similar to those in Figures 8.11 and 8.12, expressions describing this dependence can be developed. In turn, the Weibull hazard function can then be used to provide generic failure rate versus age curves using Equations 8.7 and 8.8.

APPENDIX A
PVC PRESSURE PIPES DESIGN FOR DYNAMIC STRESSES

INTRODUCTION

PVC pressure pipes are designed on the basis of a burst regression line for pipes subjected to constant internal pressure. From this long-term testing and analysis, nominal working pressure classes are allocated to pipes as a first indication of the duty for which they are suitable. However, there are many other factors that must be considered, including the effects of dynamic loading. While most gravity pressure lines operate substantially under constant pressure, pumped lines frequently do not, and it is essential that the effects of this type of loading be considered in the pipeline design phase to avoid premature failure. This appendix is intended to assist in the selection of pipe class for PVC pipes in applications involving transient and cyclic operating pressures.

Pressure fluctuations in pumped mains result from events such as pump start-up and shutdown, or valves opening and closing. The approach adopted for pipe design and class selection when considering these events, depends on the anticipated frequency of the pressure fluctuation as follows:

- For random, isolated surge events (e.g., those which result from emergency shutdowns), the designer must ensure that the maximum and minimum pressures experienced by the system are within acceptable limits.
- For frequent, repetitive pressure variations, the designer must consider the potential for fatigue and design accordingly.

DEFINITIONS

For the purposes of this appendix, surge is defined as a rapid, very short-term pressure variation caused by an accidental, unplanned event such as an emergency shutdown resulting from a power failure. Surge events are characterized by high pressure rise rates, with no time spent at the peak pressure. The maximum duration of a surge event is about five minutes.

Fatigue is associated with a large number of repetitive events. The key factors to consider are the size and frequency of the repeated event. For large pressure cycles, a lower number of events can be tolerated in the pipe lifetime. For smaller pressure changes, a greater number of events is acceptable.

The gradual diurnal pressure changes which occur in most distribution pipelines as a result of demand variation will not cause fatigue. The only design consideration required for this type of pressure fluctuation is that the maximum pressure should not exceed the pressure rating of the pipe.

SURGE DESIGN

It has long been recognized that PVC pipes are capable of handling short-term stresses far greater than the long-term loads upon which they are designed and dimensioned. However, this characteristic feature has, to date, not been utilized in design in Australia. Previous design recommendations have advised that the peak pressure should not exceed the nominal working pressure of the pipe.

Recently, studies in the UK have examined the response of plastics pipes to short-term over-pressurization. This work has culminated in the publication of a design recommendation (UK Water Industry Research Limited 2000) allowing significant over-pressurization of plastics pipes during rapid surge events. However, while these results highlight the capabilities of plastics pipes in this respect, they do not discuss the effects of over-pressurization on a pipe where it is restrained by tapping bands or flange connections. The effect on other pipeline components is also not considered. For example, it is well known that the complex geometry of pipeline fittings can amplify stress levels. Equivalent performance of plastics fittings to pipes during over-pressurization cannot be assumed.

Field tests of installed pipelines are conducted to verify the pipeline design and construction. These tests are generally carried out at a pressure of 1.25 times the maximum system operating pressure. Since the test pressure is generally the highest pressure a pipeline will experience, the field test serves to demonstrate that all system components, including items such as anchor blocks, which are formed in-situ, will be satisfactory for the operating conditions.

The demonstrated ability of plastics pipes to accommodate short-term surge pressures can be utilized to advantage by allowing some over-pressurization. However, it is recommended that the peak design pressure is limited to the field test pressure of that pipeline, in order to ensure that the pipeline as a whole is capable of performing under these conditions. Where plastics fittings are used, peak pressures should be limited to the rated working pressure of the fitting. Thus, for example, a system operating normally at 1.2 MPa, with occasional surging to 1.5 MPa, will require PN 12 pipe and PN 15 fittings, and system testing should be conducted at 1.5 MPa. However, if the installation were to be tested at 1.5 times the rated pressure of the pipe, surge pressures would still be limited to 1.5 MPa, unless higher pressure-rated fittings were used.

Where the generation of negative pressures is anticipated, the possibility of transverse buckling should be considered. This topic is well covered in other technical literature.

FATIGUE DESIGN

Many materials will fail at a lower stress when subjected to cyclic or repetitive loads than when under static loads. This type of failure is known as fatigue. For PVC pipes, fatigue is only relevant where a large number of stress cycles are anticipated. The important factors to consider are the magnitude of the stress fluctuation and the loading frequency. Where large stress fluctuations are predicted, fatigue design may be required where the total number of cycles in the operational lifetime of the pipe exceeds 100,000. For smaller stress cycles, a larger number of cycles can be tolerated.

The fatigue response of thermoplastics pipe materials, particularly PVC, has been extensively investigated. The results of laboratory studies can be used to establish a relationship between stress range, defined here as the difference between the maximum and minimum stress, and the number of cycles to failure. From these relationships, it is possible to derive load factors that can be applied to the operating pressures, to enable the selection of an appropriate class of pipe.

This type of experimental data inevitably has a degree of scatter, and it has been Australian practice, to adopt the lower bound for design purposes. This approach is retained here because it ensures the design has a positive safety factor, and recognizes that pipelines may sustain minor surface damage during installation, which could promote fatigue crack initiation. Note that for fatigue loading situations, the maximum pressure reached in the repetitive cycle should not exceed the static pressure rating of the pipe.

In common with many materials, PVC pipes exhibit a fatigue threshold. That is, below a particular amplitude in SIF, there is no cyclic fatigue effect. Thus, if the pressure amplitude in a pipeline is low, the effects of the static rather than cyclic loading determine the performance. The cyclic fatigue threshold for PVC has been determined as coinciding with the pressure amplitude causing failure at 10^6 cycles. The re-rating factor for cyclic loading of PVC pipes therefore does not diminish further for expected lifetimes beyond one million cycles.

Table A.1 gives the recommended fatigue load factors for PVC pipes. This is also graphically presented in Figure A.1.

Table A.1
Recommended fatigue load factors for PVC pipes

Total cycles	Approximate no. of cycles/day for 100-year life	PVC fatigue load factors
26,400	1	1.00*
100,000	3	1.00
500,000	14	0.62
1,000,000	27	0.50
10,000,000	274	0.50
100,000,000	2,740	0.50

* Maximum allowable factor for fatigue loading pressure cycles.

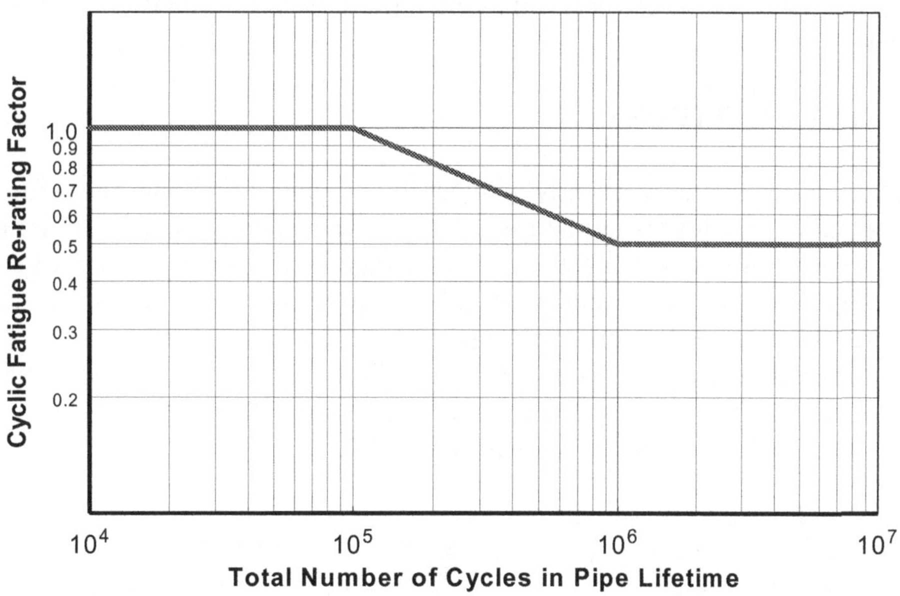

Figure A.1 Fatigue load factors for PVC pipes

PROCEDURE

To select the appropriate pipe class for fatigue loading, the following procedure should be adopted:

- Estimate the likely pressure range, i.e., the maximum pressure minus the minimum pressure.
- Estimate the frequency or the number of cycles per day which are expected to occur.
- Determine the required service life, and calculate the total number of cycles which will occur in the pipe lifetime.
- Using Table A.1 or Figure A.1, look up the fatigue load factor for PVC and the number of cycles.
- Divide the pressure range by the fatigue load factor to obtain an equivalent operating pressure.
- Use the equivalent operating pressure to determine the class of pipe required.

DEFINITION OF PRESSURE RANGE

For simplicity, the pressure range is defined as the maximum pressure minus the minimum pressure, including all transients, experienced by the system during normal operations, as shown in Figure A.2. The effect of accidental conditions such as power failure may be excluded. Figure A.2 also illustrates the definition of a cycle as a repetitive event. In some cases, the cycle pattern will be complex, and it may be necessary to also consider the contribution of secondary cycles.

COMPLEX CYCLE PATTERNS

Pumping systems are frequently subject to surging following the primary pressure transient on switching. Such pressure surging decays exponentially and, in effect, the system is subjected to a number of minor pressure cycles of reducing magnitude. In order to take this into account, the effect of each minor cycle is related to the primary cycle in terms of the number of cycles, which would produce the same crack growth as one primary cycle.

Using this technique, it is shown (Joseph 1984) that a typical exponentially decaying surge regime is equivalent to two primary cycles. Thus, for design purposes, the primary cycle amplitude only is considered, with the frequency doubled.

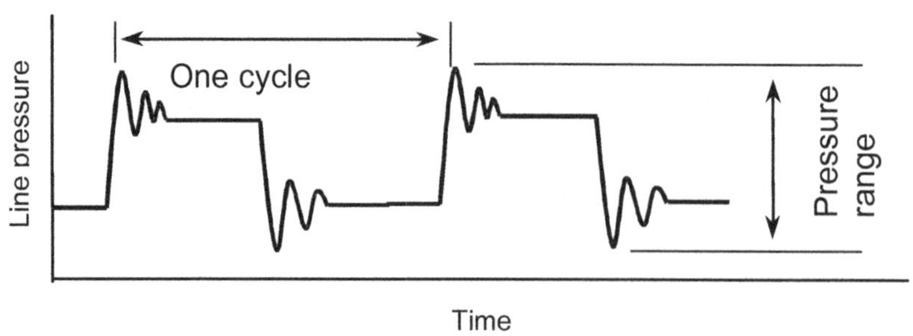

Figure A.2 Pressure cycle

In general, a similar technique may be applied to any situation where smaller cycles exist in addition to the primary cycle. Empirically, crack growth is related to stress range according to $(\Delta\sigma)^{3.2}$. Thus, n secondary cycles of magnitude $\Delta\sigma_1$ may be deemed equivalent in effect to one primary cycle, where:

$$n = \left(\frac{\Delta\sigma_0}{\Delta\sigma_1}\right)^{3.2}$$

For example, a secondary cycle of half the magnitude of the primary cycle:

$$n = \left(\frac{2}{1}\right)^{3.2} = 9.2$$

so it would require nine secondary cycles to produce the same effect as one primary cycle. If they are occurring at the same frequency, the effective frequency of primary cycling is increased by 1.1 for the purpose of design.

Example

A sewer rising main has a pump pressure, including static lift and friction losses, of 600 kPa. When the pump starts up, the pressure rises rapidly to 1.1 MPa, before decaying exponentially to the static pumping pressure.

On pump shutdown, the minimum pressure experienced by the system is 100 kPa. On average, the pump will start up 8 times per day. A minimum life of 100 years is required.

The maximum pressure experienced indicates that a minimum class of PN 12 (1.2-MPa rated pressure) will be required.

A fatigue analysis is now needed in order to determine the suitability or otherwise of PN 12.

In this system, the pressure range is 1 MPa. The pump will start up approximately 292,000 times in a 100-year lifetime. However, the exponential cycle pattern means that this should be doubled for design purposes. Therefore, the system should be designed to withstand approximately 584,000 cycles in a 100-year lifetime.

Using Figure A.1 to determine the fatigue load factors for PVC pipes at 5.8×10^5 cycles, gives the following class selection:

- Material PVC.
- Load factor (Figure A.1) 0.6.
- Equivalent operating pressure (MPa) 1.0/0.6 = 1.67.
- Minimum pipe class selection PN 18.

The fatigue analysis thus determines that although PN 12 is adequate for the total pressure, PN 18 pipe is needed in order to cope with fatigue effects.

Effect of Temperature

Few data have been established concerning the relationships between fatigue performance and temperature. The literature suggests that effects are marginal, if any, within the range of normal pipeline operating temperatures. It might be expected that higher temperatures could actually inhibit fatigue crack growth, by blunting of the crack tip through the reduced yield strength.

Normal temperature de-rating principles should be applied in class selection for maximum operating pressure. Fatigue should then be considered separately, and the highest pressure class selected, i.e., select the highest class arrived via:

- Static design including temperature derating.
- Dynamic design as covered herein.

Safety Factors

The fatigue design guidelines presented in Table A.1 are based on the lower bound of laboratory test data. They are therefore considered conservative, and no additional safety factor need be applied in general. However, where the magnitude or frequency of dynamic stresses cannot be estimated in design with any reasonable degree of accuracy, appropriate caution should obviously be applied. This judgment is in the hands of the designer.

While it is always possible to predict the steady operating conditions with good accuracy, it will occasionally be the case, in complex systems, that it is impossible to predict the extent of surge pressures. In such circumstances, relatively low-cost surge mitigation techniques (e.g., solid-state, soft-start motor controllers) should be considered. It is, of course, recommended that actual operating conditions for all systems should be measured, as a matter of routine, when the system is commissioned. Should surge pressure amplitudes in the event exceed expected levels, it is relatively easy to retrofit control equipment to ensure that they are kept in check.

Design Hints

To reduce the effect of dynamic fatigue in an installation, the designer can:

- Limit the number of cycles by:
  - Increasing well capacity for a sewer pumping station.
  - Matching pump performance to tank size to eliminate short demand cycles for an automatic pressure unit.
  - Using double-acting float valves or limiting starts on the pump by the use of a time clock when filling a reservoir.
- Reduce the dynamic range by:
  - Eliminating excessive water hammer.
  - Using a larger bore pipe to reduce friction losses.

Fittings

Thermoplastics fittings present a special problem. The geometry of fittings can result in complex stress patterns, which "amplify" the apparent stress cycle. An apparently harmless pressure cycle can thus produce a damaging stress cycle, leading to a shortened fatigue life. This is particularly important in the case of branch fittings such as tees. In addition, the situation can be aggravated further by the existence of stresses from other sources, for example, bending stresses induced by flexing under hydraulic thrust in improperly supported systems.

Since the design of fittings is not standardized, the fittings manufacturer should be consulted for recommended de-rating factors for cyclic loading conditions. It is therefore necessary to consider fittings separately from pipe.

APPENDIX B
OPTICAL METHOD FOR ASSESSING THE RESIN DISPERSION IN PVC PIPES

SCOPE

This procedure describes a method for assessing the level of dispersion of the resin in PVC pipes. The method is applicable to PVC-U, PVC-M, and PVC-O pipes for pressure applications.

PURPOSE

It has been established that the toughness of PVC, as demonstrated by resistance to slow crack growth, is dependant upon the level of gelation. The extent to which residual PVC grain structure is evident in a pipe wall is one of the indicators of the level of gelation.

Note: The method can be applied to samples taken as the extrudate exits the die at the start of a production run. Assessment of the dispersion can therefore be completed before any pipe is actually produced.

APPARATUS

1.1. Microtome – a rotary microtome capable of cutting sections of PVC to10–20 μm thick.

1.2. A microscope having a magnification in the range of 32–40×. Both transmitted and incident light sources are permitted for viewing the specimens.

PREPARATION OF SPECIMENS

1.3. Cut a small piece of pipe suitable for clamping in the microtome. Ensure the sample includes the full wall thickness of the pipe.

1.4. Mount the piece of pipe in the microtome and clamp firmly.

1.5. Proceed to microtome sections approximately 10–20 μm thick from the piece of pipe. These sections shall be cut through the full wall thickness of the pipe. They may be cut in either the direction of extrusion, or at right angles to the extrusion direction.

Note: While the assessment can equally be performed on sections cut in either direction, it is preferable to standardize on one direction for routine assessments.

1.6. Take a complete section across the full wall thickness and place on a microscope slide. Cover with another slide or cover slip.

EXAMINATION

1.7. Place the specimen on the microscope stage and illuminate with either incident or transmitted light.

1.8. Examine the specimen at a magnification in the range of 32–40×. If the section is too large to fit in one field of view, locate the area of poorest dispersion.

1.9. Compare the dispersion of the microtomed section with the standard photographs showing poor and complete dispersion, and apply a rating.

INTERPRETATION

1.10. A pipe with dispersion equal to a photograph of the poor sample is considered to be under-processed.

1.11. Specimens with a dispersion rating between "poor" and "complete" shall be rated on the basis of a correlation, established by in-house experience, between dispersion and mechanical properties such as impact resistance or fracture toughness.

Note: A dispersion rating poorer than "complete" does not mean the pipe is not fit for the purpose or will fail any of the product Standard tests. However, as a sample approaches the "poor" end of the scale, it should be viewed as a warning sign that more detailed examination of the pipe quality is warranted.

REPORT

The following information shall be reported:

1.12. Nominal diameter, DN.

1.13. Pressure classification, PN.

1.14. Date of manufacture.

1.15. Material (PVC-U, PVC-M, or PVC-O).

1.16. Dispersion rating in accordance with the "in-house" procedure.

STANDARD PHOTOGRAPHS   DISPERSION IMAGES

Complete Dispersion

Transmitted light

Reflected light

Poor Dispersion

Transmitted light

Reflected light

Figure B.1 Microtomed samples cut perpendicular to the direction of material flow

Transmitted light *Reflected light*

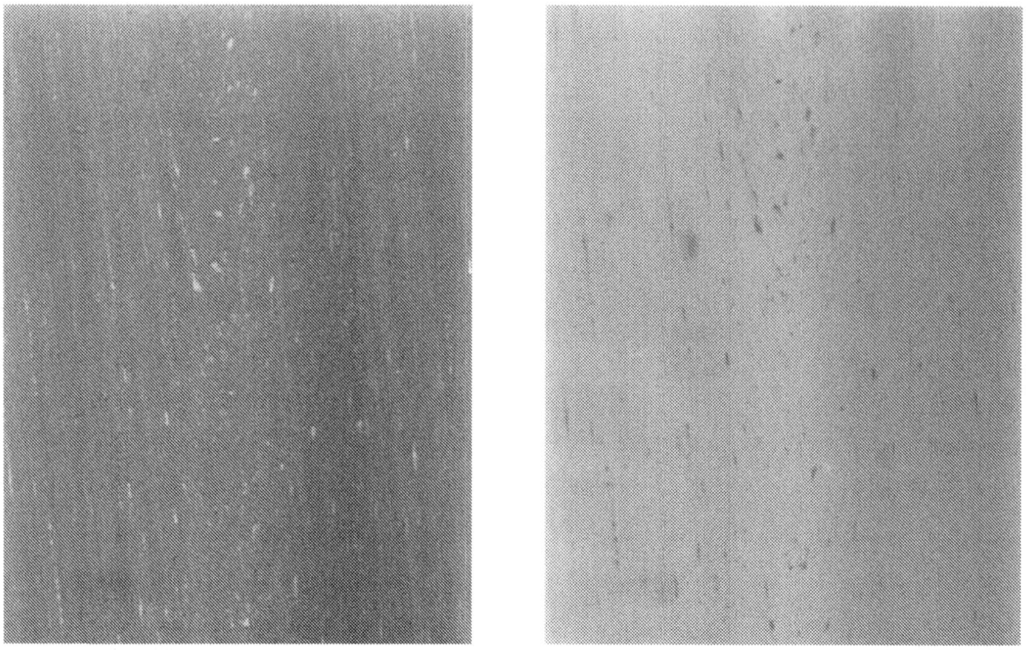

Figure B.2. Microtomed samples cut through the thickness of the pipe wall, and parallel to the direction of material flow

APPENDIX C
PROGRAM LISTING FOR UKWIR PVC_U FAILURERATES.PY

The calculation of average failure rates vs. age curves for UK PVC- pipes in Chapter 5 has been implemented using the PYTHON 2.3 programming language. PYTHON is freeware, and can be installed from the internet (www.python.org). The full listing for the program UKWIR PVC_U FailureRates.py is given below. Text that is proceeded by "#" or "#*" is descriptive, and not part of the PYTHON code.

```
#**********************START OF PROGRAM****************************************

import os
import sys
import string
import math

_root = 'O://AWWARF PVC project//PVC Field performance//Python Scripts for failure rates'

def readPipeData(inpfname,dict):
   inpData  = file(inpfname)
   Query = inpData.readlines()
   inpData.close()
   for qrow in Query[1:]:
      if qrow != '':
         try:
            qrow = qrow.replace('\n','')
            qrow = qrow.replace(' ','')
            try:
               var = string.split(qrow,',')
               id = int(var[0])
               installYr = int(var[4])
               length = float(var[7])
            except:
               pass
            dict[(id)] = [installYr,length]
         except:
            pass
      if qrow == '':
         pass
   return dict

pipeDict = {}
readPipeData(_root+'//PVC-U failures (UKWIR database)(Mains)(Fracture failures Only).csv',pipeDict)

failureYears                                                                                          =
[0,1980,1981,1985,1987,1988,1990,1991,1992,1993,1994,1995,1996,1997,1998,1999,2000,2001,2002,2
003]
```

```
#********************CALCULATE EXPOSURE**********************#

exposure = {}
for id in pipeDict.keys():
   iYear = pipeDict[id][0]
   length = pipeDict[id][1]
   for year in failureYears:
      age = year - iYear
      try:
         exposure[age] += length
      except:
         exposure[age] = length


def readBurstData(inpfname,dict):
   inpData  = file(inpfname)
   Query = inpData.readlines()
   inpData.close()
   for qrow in Query[1:]:
      if qrow != '':
         try:
            qrow = qrow.replace('\n','')
            qrow = qrow.replace(' ','')
            try:
               var = string.split(qrow,',')
               count = int(var[10])
               age = int(var[11])
            except:
               pass
            dict[(count)] = age
         except:
            pass
      if qrow == '':
         pass
   return dict

burstDict = {}
readBurstData(_root+'//PVC-U failures (UKWIR database)(Bursts)(Allpipes)(Fractures Only).csv',burstDict)

#********************CALCULATE FAILURE AGES*******************#

failureTimes = {}
for id in burstDict.keys():
   t = int(burstDict[id])
   try:
      failureTimes[t] += 1
   except:
      failureTimes[t] = 1
```

```
#******************CALCULATE FAILURE RATES*********************#

failureRates = {}
for t in failureTimes.keys():
    try:
        failureRates[t] = float(failureTimes[t]) / float(exposure[t])
    except:
        pass

def wrtFailRate(fname,dict1,dict2):
    data = file(fname,'w')
    line = "%s" %(' ')
    line += "\nPipe Age (years), Expected failure rate (per 100km/per year), # of observations, maxFailRate, minFailRate, +error, -error"
    for id in dict1.keys():
        line += "\n%s," %id
        failRate = float(dict1[id]*100e3)
        line += "%s," %failRate
        n = int(dict2[id])
        line += "%s," %n
        maxFR = failRate*math.exp(2.0/math.sqrt(n))
        minFR = failRate/math.exp(2.0/math.sqrt(n))
        line += "%s," %maxFR
        line += "%s," %minFR
        plusError = maxFR - failRate
        negError = failRate - minFR
        line += "%s," %plusError
        line += "%s" %negError
    data.write(line)
    data.close()


wrtFailRate(_root+'//PVC-U failureRate vs Age(AllPipes)(Fractures Only).txt',failureRates,failureTimes)
```

APPENDIX D
PROGRAM LISTING FOR PVCLIFETIME.PY

The Monte Carlo simulation for buried PVC-U pipes has been implemented using the PYTHON 2.3 programming language. PYTHON is freeware, and can be installed from the internet (www.python.org). For this particular study, PYTHON has allowed complex failure models to be combined with statistical analysis, While maintaining an emphasis on the physical aspects of the problem rather than computing science issues. The full listing for the program PVCLifetime.py is given below. Text that is proceeded by "#" or "#*" is descriptive, and not part of the PYTHON code.

```
#**********************START OF PROGRAM*****************************************

import math
import Weibull
import random
import os
import sys
import string

_root = 'O://PythonPVCOutput'  # Commment: Sets directory in which output files are written

#***SECTION 1 - FUNCTION DEFINITIONS FOR FRACTURE MECHANICS MODEL***********

def getSip(p,ro,ri):  # Comment: circumferential stress from internal pressure
   val = p*(ro**2+ri**2)/(ro**2-ri**2)
   return val

def getI(b):  # Comment: Second moment of area for pipe
   val = (b**3)/12
   return val

def getE(E0,t,c):  # Comment: Time dependent Young's Modulus
   val = (E0*t**(-c))
   return val

def getDeflect(E,I,Dm,Esoil,gamma,H, Psurf):  # Comment: Diametrical deflection
   val = 0.0001*Dm*((gamma*H)+Psurf)/((8*E*I/Dm**3)+0.061*Esoil)
   return val

def getSid(E,b,deflect,Dm):  # Comment: Circumferential stress from diametrical deflection
   val = 2*E*b*deflect/(0.1488*(Dm**2))
   return val

def incadef(D,Kapp,m):  # Comment: Incremental crack growth under applied SIF
   val = (31.54e6*D*(Kapp**m))
   return val
```

```
def getb(ro,ri): # Comment: Pipe wall thickness
    val = ro-ri
    return val

def getDm(ri,b): # Comment: Pipe mean diameter
    val = 2*(ri+b/2)
    return val

def getYp(adef,b): # Comment: LEFM geometry factor for internal pressure
    val = 1.824-4.669*(adef/b)+(14.959*(adef/b)**2)-(6.341*(adef/b)**3)
    return val

def getYd(adef,b): # Comment: LEFM geometry factor for diametrical deflection
    val = 0.803+0.353*(adef/b)-(1.640*(adef/b)**2)-(1.003*(adef/b)**3)
    return val

def getYr(adef,b): # Comment: LEFM geometry factor for residual stress
    val = 0.727+2.791*(adef/b)-(2.214*(adef/b)**2)-(2.149*(adef/b)**3)
    return val

def getKp(Sip,Yp,adef): # Comment: LEFM applied SIF for internal pressure
    val = Sip*Yp*(3.14159*adef)**0.5
    return val

def getKd(Sid,Yd,adef): # Comment: LEFM applied SIF for diametrical deflection
    val = Sid*Yd*(3.14159*adef)**0.5
    return val

def getKr(Sir,Yr,adef): # Comment: LEFM applied SIF for residual stress
    val = Sir*Yr*(3.14159*adef)**0.5
    return val

def getKapp(Kp,Kd,Kr): # Comment: Total applied SIF
    val = Kp+Kd+Kr
    return val

def getIncubTime(CCOD,E0, sigma0, KappInit, c, q): # Comment: Incubation time
    val = (E0*sigma0*CCOD/(KappInit**2))**(1/(c+q))
    return val

def getSigmaY(sigma0,t,q): # Comment: Time-dependent Yield stress
    val = sigma0*(t**(-q))
    return val

def getSigmaNS(sigmaY,adef,b): # Comment: Net section stress
    val = sigmaY*(1-(adef/b))
    return val

def getSapp(Sip,Sid,Sir): # Comment: Total applied stress
    val = Sip+Sid+Sir
    return val
```

```
def getYtot(Yp,Sip,Yd,Sid,Yr,Sir): # Comment: LEFM geometry factor for combined loading
   val = (Yp*Sip+Yd*Sid+Yr*Sir)/(Sip+Sid+Sir)
   return val

def getSigmaFrac(Kic,Ytot,adef): # Comment: Stress required for brittle fracture
   val = Kic/(Ytot*((3.14159*adef)**0.5))
   return val

#******END OF SECTION 1*******************************************************

#***SECTION 2 – INPUT VARIABLES**********************************************

input = {}
input['Kic'] = 5.0        # Comment: Fracture toughness of pipe material (Mpa m^0.5)
input['D'] = 7.19e-13        # Comment: Slow crack growth parameter for PVC-US (Tin stabilised)
input['m'] = 17.8        # Comment: Slow crack growth power exponent for PVC US (Tin stabilised)
input['p'] = 0.95    # Comment: Internal Pressure (MPa)
input['ro'] = 6.10E-02        # Comment: Inner radius (Metres)
input['ri'] = 5.39E-02          # Comment: Outer radius (Metres)
input['E0'] = 1775           # Comment: Short term Young's modulus (MPa) (based on 1 year)
input['c'] = 0.05              # Comment: Viscoelastic parameter for Young's modulus reduction over time
(Mod in MPa, time in years)
input['Esoil'] = 1.0        # Comment: Soil modulus (MPa)
input['gamma'] = 2.02          # Comment: Soil unit weight (KN/m^3)
input['H'] = 0.4            # Comment: Soil cover depth (Metres)
input['Psurf'] = 15.0         # Comment: Surface loading (KPa)
input['Sir'] = 3.16          # Comment: Residual hoop stress at pipe inner surface (MPa)
input['adef_shape'] = 1.621   # Comment: Weibull shape parameter for initial defect size
input['adef_scale'] = 4e-4   # Comment: Weibull scale parameter for initial defect size
input['sigma0'] = 29.25       # Comment: Short term yield stress (MPa) (based on 1 year)
input['q'] = 0.04           # Comment: Viscoplastic parameter for yield stress reduction over time (stress in
MPa, time in years)
input['N'] = 100    # Comment: Number of pipe segments


#***********************END OF SECTION 2**************************************

#************SECTION 3 - MONTE CARLO SIMULATION OF PIPE LIFETIMES*********

def getLifetimes(input):
   T = 5000  # Comment: Time period for simulation
   time = range(1,T)
   adef_shape = input['adef_shape'] # Comment: assigns Weibull shape parameter for defect distribution
   adef_scale = input['adef_scale'] # Comment: assigns Weibull scale parameter for defect distribution
   ro = input['ro']
   ri = input['ri']
   p = input['p']
   E0 = input['E0']
   c = input['c']
   Esoil = input['Esoil']
   gamma = input['gamma']
```

```
H = input['H']
Psurf = input['Psurf']
Sir = input['Sir']
N = input['N']
Kic = input['Kic']
D = input['D']
m = input['m']
sigma0 = input['sigma0']
q = input['q']
Sip = getSip(p,ro,ri) # Comment: Calculates internal pressure for input variables #
b = getb(ro,ri)
Dm = getDm(ri,b)
I = getI(b)
pipes = range(N) # Comment: Number of pipe segments in simulation
Failed = {}         # Comment: Set up dictionary for all pipes
Replaced = {}       # Comment: Set up dictionary for replaced pipes
NewPipeLog = {}  # Comment: Set up dictionary to record new pipe segments installed over time

for id in pipes:
   Failed[id] = [0]   # Comment: Initialise dictionary of failed pipes
   Replaced[id] = [0]   # Comment: Initialise dictionary of replaced pipes

for t in time:
      NewPipeLog[t] = [] # Comment: Initialise list of new pipe installation years

for i in pipes:
   adef = random.weibullvariate(adef_scale,adef_shape) # Comment: assign inherent defect sizes to all
pipe segments
   for t in time:
      if Replaced[i]:
         adef = random.weibullvariate(adef_scale,adef_shape)   # Comment: assign new initial defect in
replaced pipe
      if not Failed[i] or Replaced[i]: # Comment: check if pipe has not failed yet or has been replaced
         Yp = getYp(adef,b)
         Yd = getYd(adef,b)
         Yr = getYr(adef,b)
         Kp = getKp(Sip,Yp,adef)
         E = getE(E0,t,c)
         deflect = getDeflect(E,I,Dm,Esoil,gamma,H, Psurf)
         Sid = getSid(E,b,deflect,Dm)
         Kd = getKd(Sid,Yd,adef)
         Kr = getKr(Sir,Yr,adef)
         Kapp = getKapp(Kp,Kd,Kr)
         adef = adef + incadef(D,Kapp,m)
         sigmaY = getSigmaY(sigma0,t,q)
         sigmaNS = getSigmaNS(sigmaY,adef,b)
         Ytot = getYtot(Yp,Sip,Yd,Sid,Yr,Sir)
         sigmaFrac = getSigmaFrac(Kic,Ytot,adef)
         Sapp = getSapp(Sip,Sid,Sir)
         if sigmaFrac <= Sapp: # Comment: Check for brittle fracture failure
            Failed[i] += [t]  # Comment: Add failure year to list for pipe "i"
```

```
        Replaced[i] += [t] # Comment: Add installation year to list for  pipe "i"
        NewPipeLog[t] += [1] # Comment: Add "1" into new pipe list at time "t"

      else:
        if adef >= b: # Comment: Check for slow crack growth failure
              Failed[i] += [t]
              Replaced[i] += [t]
              NewPipeLog[t] += [1]
        else:
          if sigmaNS <= Sapp: # Comment: Check for ductile fracture failure  #
                    Failed[i] += [t]
                    Replaced[i] += [t]
                    NewPipeLog[t] += [1]

#***********************END OF SECTION 3 *****************************************

#***********************SECTION 4 - WRITE OUTPUT TO TEXT FILES******************

  wrtDict(_root+'//FailHistory.txt','Segment id,[Failure years]',Replaced)
  wrtDict2(_root+'//NewSegment.txt','Year, Number of new segments installed',NewPipeLog)

  r_dict = {}
  r_dict['replaced'] = Replaced
  r_dict['failed'] = Failed
  r_dict['newPipeLog'] = NewPipeLog

  return r_dict

def wrtDict(fname,title,dict): # Comment: Function writing failure years to each pipe segment in the pipe
dictionary
  data  = file(fname,'w')
  data.write(title)
  for id in dict.keys():
     line = "\n%s," %id
     line += "%s" %dict[id]
     data.write(line)
  data.close()

def wrtDict2(fname,title,dict): # Comment: Function writing number of new installations to the pipe
dictionary
  data  = file(fname,'w')
  data.write(title)
  for id in dict.keys():
     line = "\n%s," %id
     line += "%s" %len(dict[id])
     data.write(line)
  data.close()

#***********************END OF SECTION 4******************************************
#***********************END OF PROGRAM********************************************
```

Output from PVCLifetime.py is contained in two .txt files. FailHistory.txt lists pipe segment number in the first column, and the years when failures were recorded for that segment in the second column. Figure D.1 shows the output for the 16 segments (labeled 0 to 9). The numbers in square brackets are the failure years for each segment. For example, segment id #6 was installed originally at year 0 and failed for the first time at year 16. It was ten replaced with a new segment which failed in year 149.

The second output file, NewSegment.txt, lists the simulation year in the first column, and the number of new pipe segments installed in each year in the second column (Figure D.2).

As shown, for years 16 and 22, one new segment was installed. Referring back to Figure D.1, these new installations correspond to segment ids #6 and #9 respectively.

```
Segment id,[Failure years]
0,[0, 151]
1,[0]
2,[0]
3,[0]
4,[0]
5,[0]
6,[0, 16, 149]
7,[0]
8,[0]
9,[0, 22]
```

Figure D.1 Example of output file FailHistory.txt from PVCLifetime.py

```
Year, Number of new segments installed
1,0
2,1
3,0
4,1
5,0
6,0
7,1
8,0
9,0
10,0
11,0
12,0
13,0
14,0
15,0
16,1
17,0
18,0
19,0
20,0
21,0
22,1
```

Figure D.2 Example of output file NewSegment.txt from PVCLifetime.py

APPENDIX E
EXTRAPOLATED AVERAGE FAILURE RATE VS. AGE CURVES FOR LARGER PIPES

The Following Figures show expected average failure rate vs age curves for:

- Class 150 psi PVC-U (Tin) (USA) pipes, DN 4 inch to DN 12 inch
- Class 1.2 MPa, Series 2 PVC-U (Ca/Zn) (AUS) pipes, DN 100 mm to DN 300 mm

In each Figure, curves for three different operating pressures are shown.

CLASS 150 PSI, PVC-U (TIN) (USA) DN 4 INCHES TO DN 12 INCHES

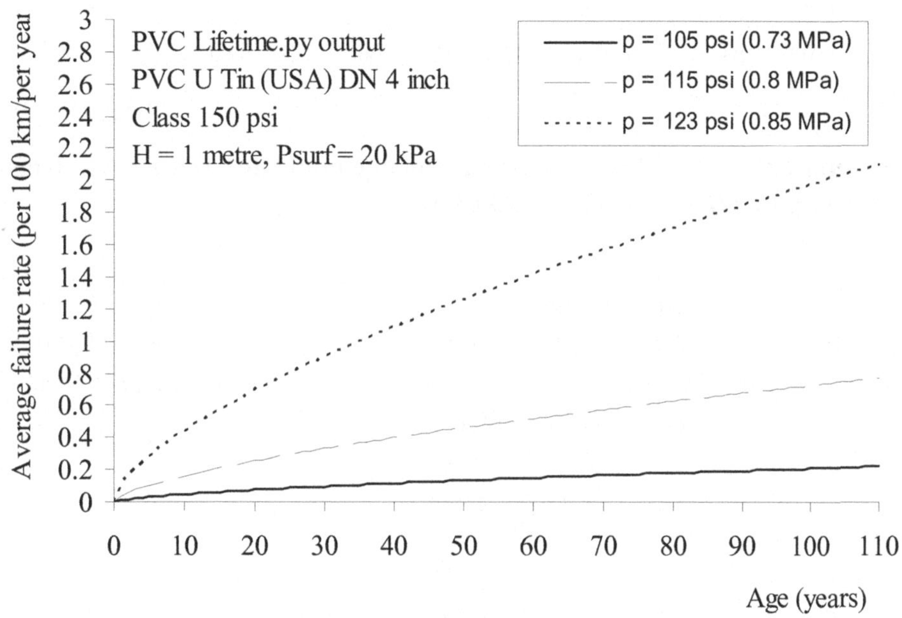

Figure E.1 Simulated average failure rate versus age curve for DN 4 inch, Class 150 psi, PVC-U Tin (USA) pipe (H = 1 m, P_{surf} = 20 kPa)

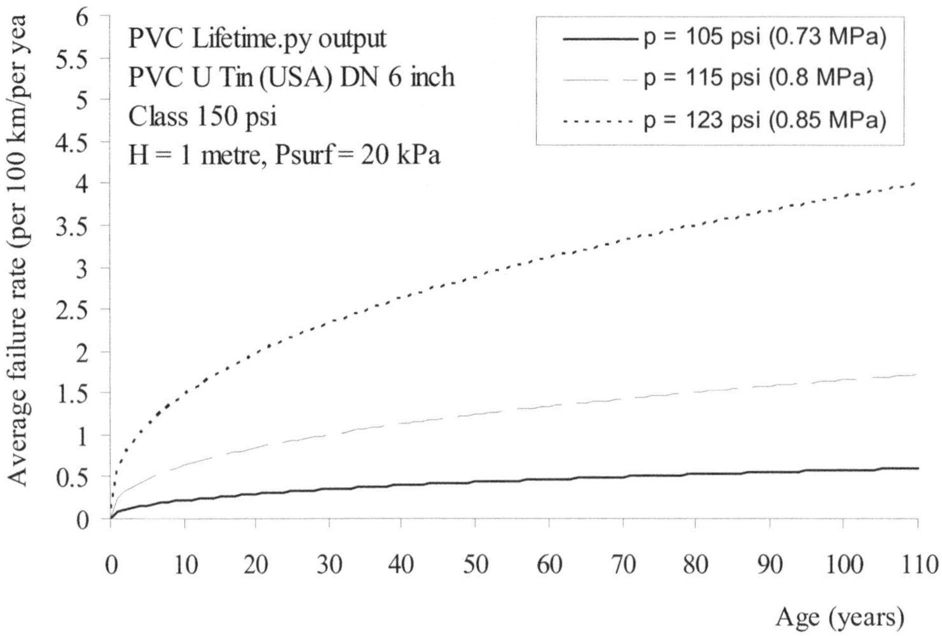

Figure E.2 Simulated average failure rate versus age curve for DN 6 inch, Class 150 psi, PVC-U Tin (USA) pipe (H = 1 m, P_{surf} = 20 kPa)

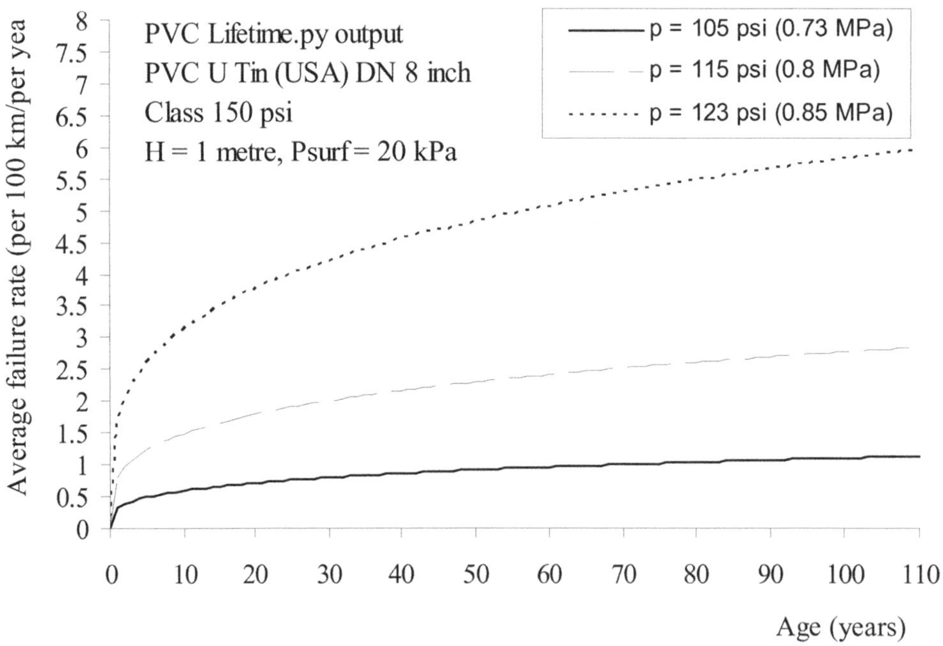

Figure E.3 Simulated average failure rate versus age curve for DN 8 inch, Class 150 psi, PVC-U Tin (USA) pipe (H = 1 m, P_{surf} = 20 kPa)

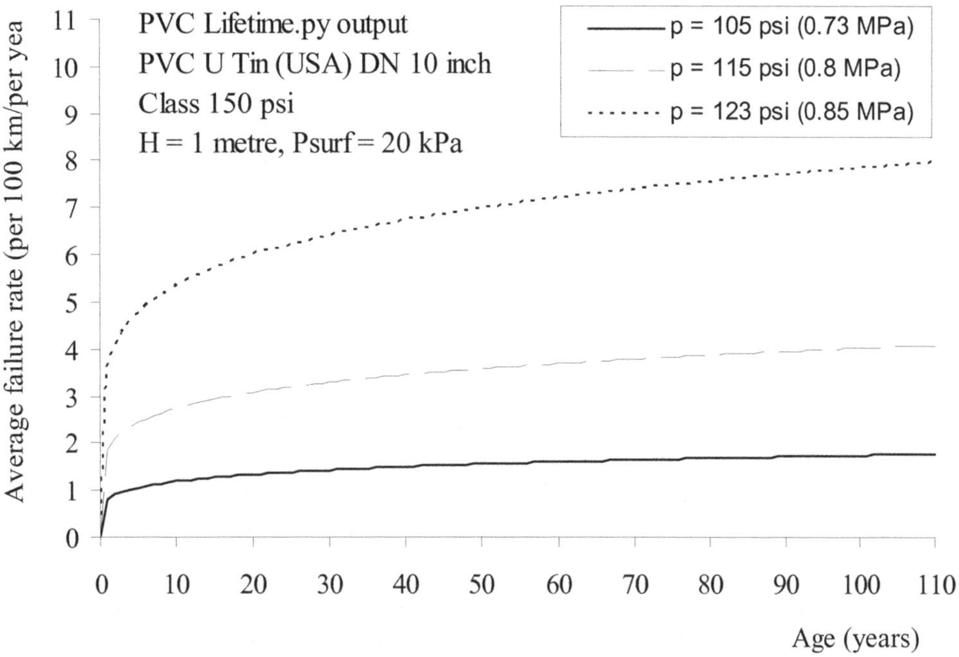

Figure E.4 Simulated average failure rate versus age curve for DN 10 inch, Class 150 psi, PVC-U Tin (USA) pipe (H = 1 m, P_{surf} = 20 kPa)

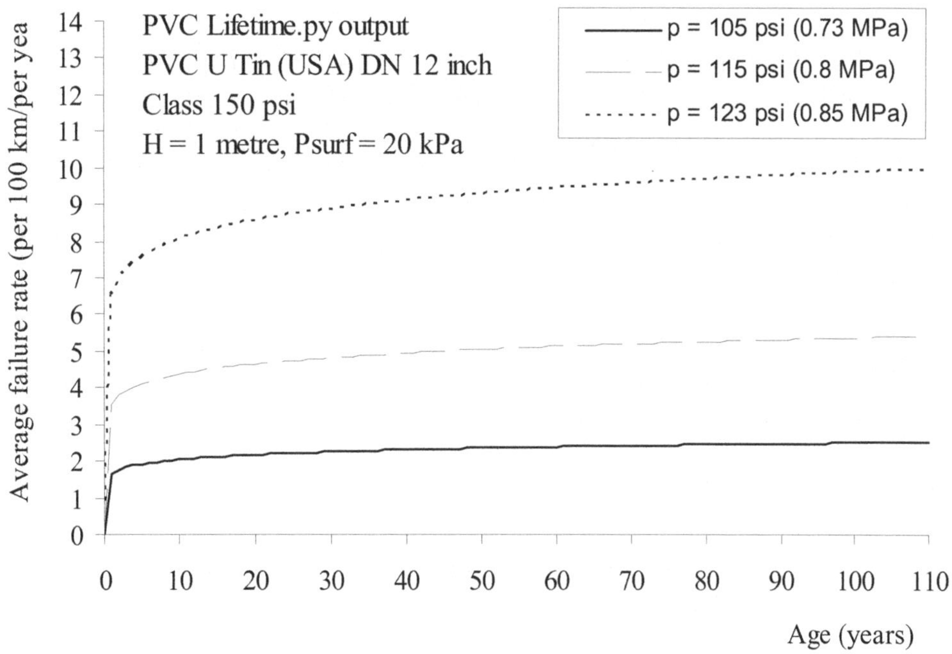

Figure E.5 Simulated average failure rate versus age curve for DN 12 inch, Class 150 psi, PVC-U Tin (USA) pipe (H = 1 m, P_{surf} = 20 kPa)

CLASS 1.2 MPA, SERIES 2 PVC-U (CA/ZN) (AUS) DN 100 MM TO DN 300 MM

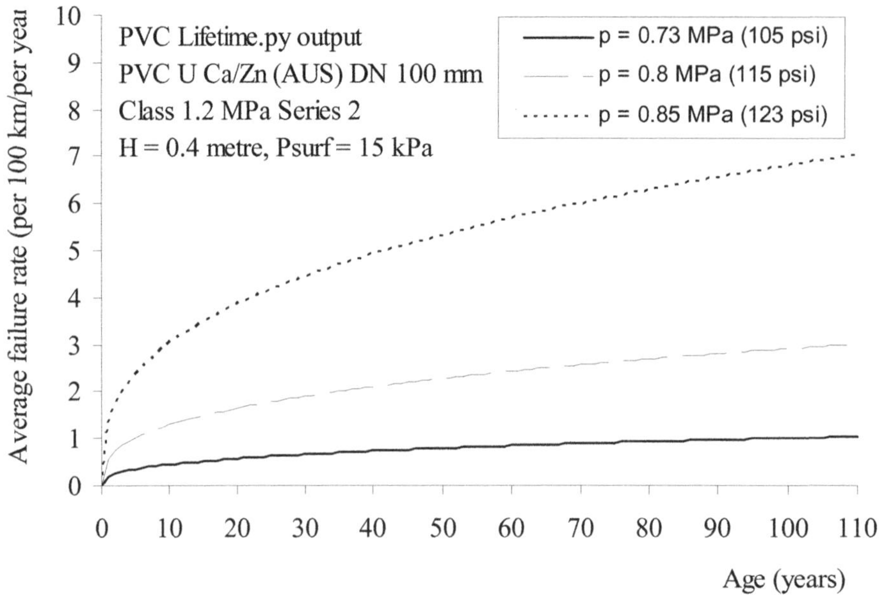

Figure E.6 Simulated average failure rate versus age curve for DN 100 mm, Class 1.2 MPa Series 2, PVC-U Ca/Zn (AUS) pipe (H = 0.4 m, P_{surf} = 15 kPa)

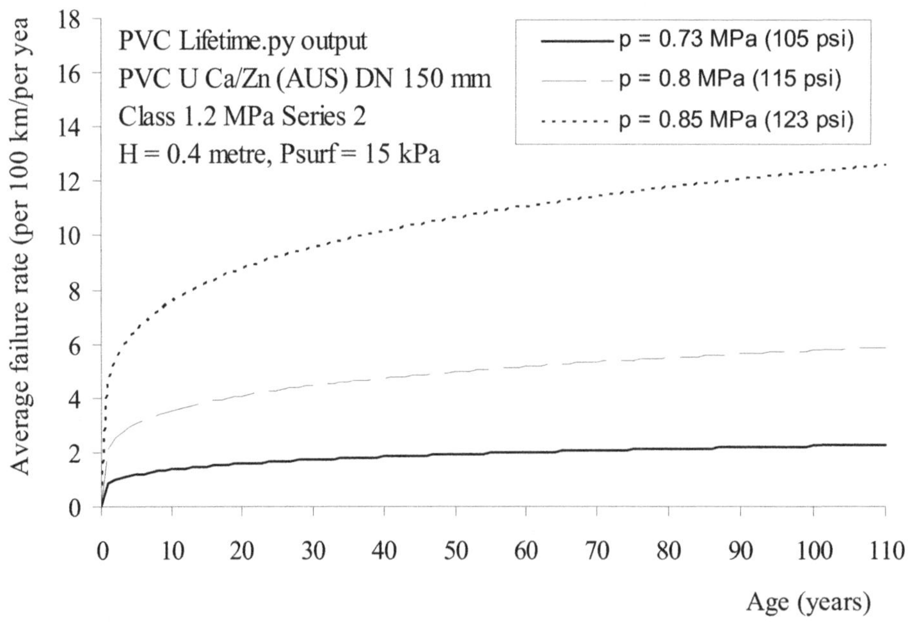

Figure E.7 Simulated average failure rate versus age curve for DN 150 mm, Class 1.2 MPa Series 2, PVC-U Ca/Zn (AUS) pipe (H = 0.4 m, P_{surf} = 15 kPa)

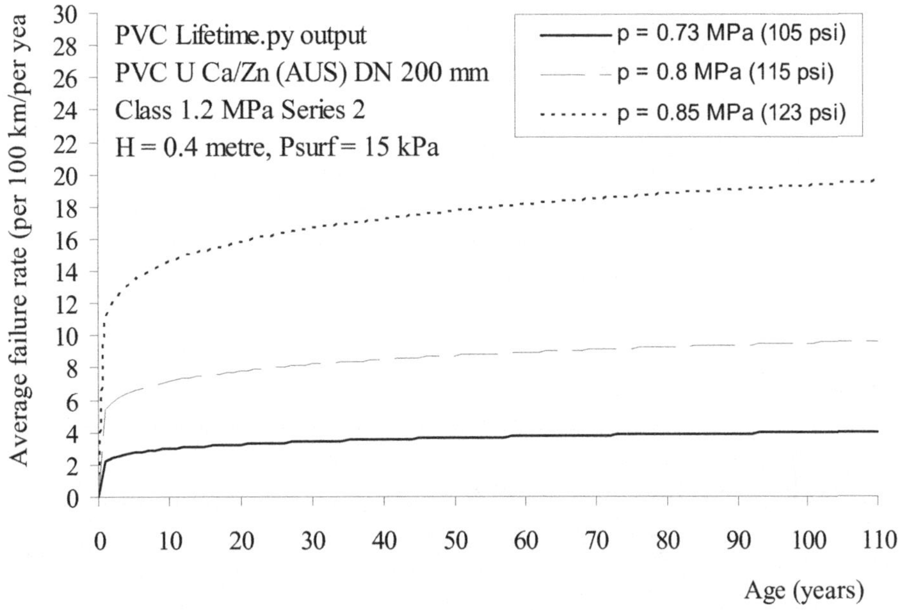

Figure E.8 Simulated average failure rate versus age curve for DN 200 mm, Class 1.2 MPa Series 2, PVC-U Ca/Zn (AUS) pipe (H = 0.4 m, P_{surf} = 15 kPa)

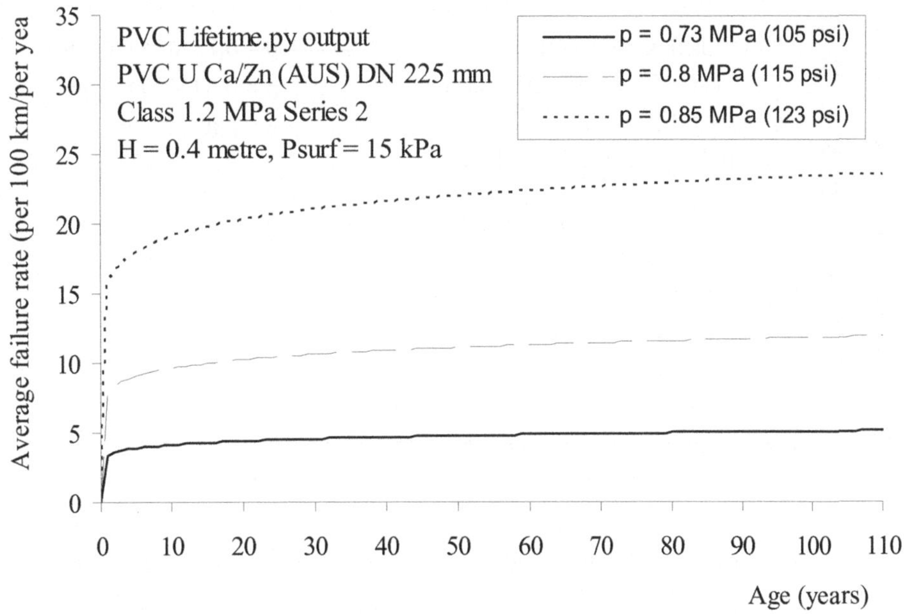

Figure E.9 Simulated average failure rate versus age curve for DN 225 mm, Class 1.2 MPa Series 2, PVC-U Ca/Zn (AUS) pipe (H = 0.4 m, P_{surf} = 15 kPa)

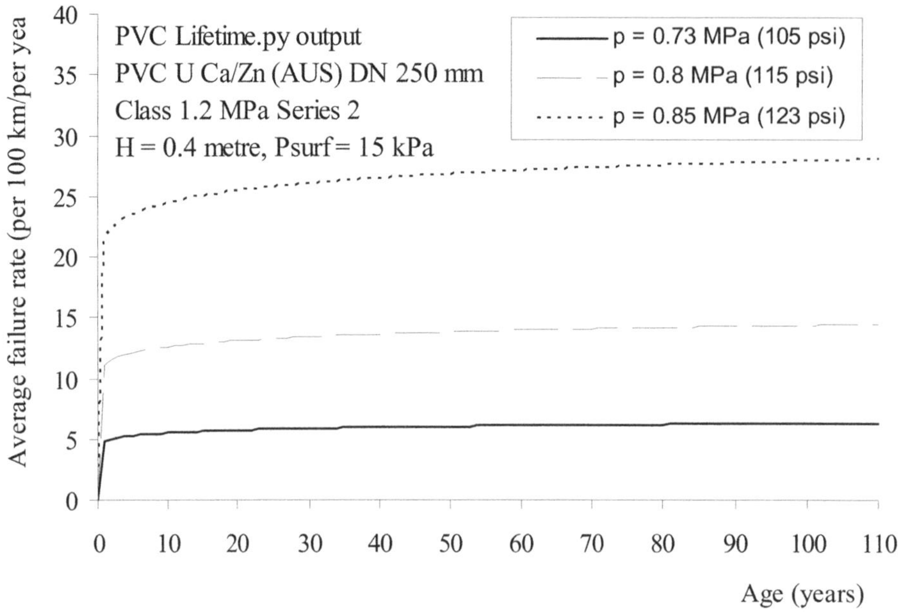

Figure E.10 Simulated average failure rate versus age curve for DN 250 mm, Class 1.2 MPa Series 2, PVC-U Ca/Zn (AUS) pipe ($H = 0.4$ m, $P_{surf} = 15$ kPa)

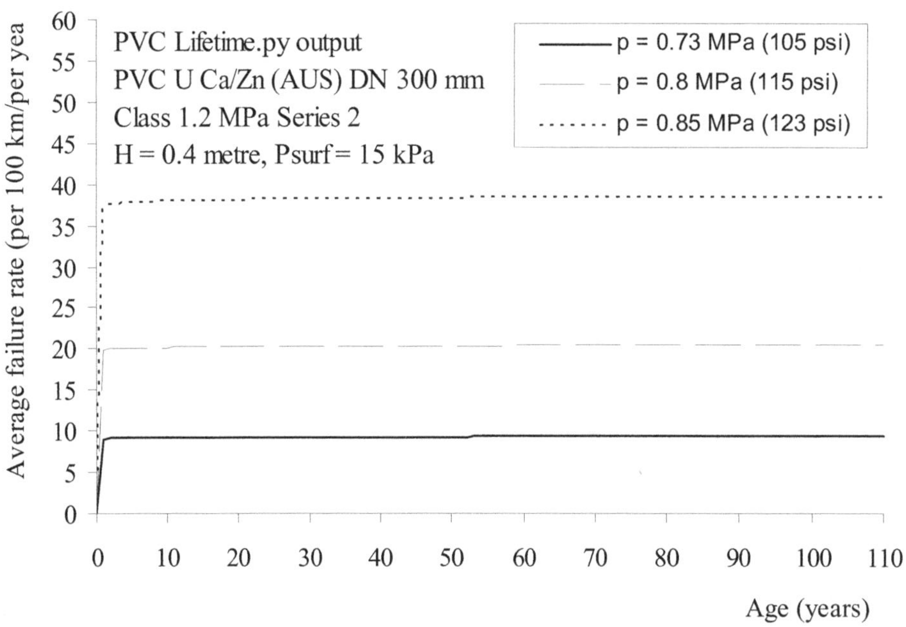

Figure E.11 Simulated average failure rate versus age curve for DN 300 mm, Class 1.2 MPa Series 2, PVC-U Ca/Zn (AUS) pipe ($H = 0.4$ m, $P_{surf} = 15$ kPa)

REFERENCES

Ahammed, M. and R.E. Melchers. 1997. Probabilistic Analysis of Underground Pipelines Subject to Combined Stress and Corrosion. *Engng. Struct.* 19(12):988-994

Alferink, F., L.-E. Janson and L. Holloway. 1997. Old unplasticised poly(vinyl chloride) water pressure pipes Investigation into Design and Durability. Plast. Rubber Compos. Process Appl. 26(2):55-58

Al–Malack, M.H. 2001. Migration of Lead from Unplasticised Polyvinyl Chloride Pipes. *Jour. Haz. Mat.*, B82:263–274.

Amer, A.R., and J.S. Shapiro. 1980. *Jour. Macromol. Sci. Chem., Part A*, 14:185.

ANSI/NSF (American National Standards Institute and NSF International). 2001. *Drinking Water System Components*, ANSI/NSF 14. Ann Harbor, Michigan: American National Standards Institute.

ANSI/NSF (American National Standards Institute and NSF International). 2002. *Drinking Water System Components – Health Effects*, ANSI/NSF 61. Ann Harbor, Michigan: American National Standards Institute.

ANSYS Inc., 1999. *ANSYS Theory Reference, Release 5.6,* ANSYS Inc., Southpointe, 275 Technology Drive, Canonsburg, PA 15317, USA, November 1999.

ASTM (American Society for Testing and Materials). 1992. *Standard Practice for Oven Heat Stability of PVC Compositions*, ASTM D2115. Philadelphia, Pa.: American Society for Testing and Materials.

ASTM (American Society for Testing and Materials). 1998. *Standard Specification for Oriented Poly(Vinyl Chloride), PVCO, Pressure Pipe*, F1483-98. Philadelphia, Pa.: American Society for Testing and Materials.

ASTM (American Society for Testing and Materials). 2002. *Standard Specification for Rigid Poly (Vinyl Chloride) (PVC) Exterior-Profile Extrusions Used for Assembled Windows and Doors*, ASTM D4726–92. Philadelphia, Pa.: American Society for Testing and Materials.

ASTM (American Society for Testing and Materials). 2002. *Standard Practice for Calculation of Color Tolerances and Color Differences From Instrumentally Measured Color Coordinates*, ASTM D2244-02. Philadelphia, Pa.: American Society for Testing and Materials.

ASTM (American Society for Testing and Materials). 2003: *Standard Specification for Rigid Poly (Vinyl Chloride) (PVC) and Related PVC and Chlorinated Poly (Vinyl Chloride) (CPVC) Building Products Compounds*. ASTM D4216. Philadelphia, Pa.: *American Society for Testing and Materials*.

ASTM (American Society for Testing and Materials). 2004. *Standard Specification for Poly(Vinyl Chloride) (PVC) Pressure-Rated Pipe (SDR Series)*, ASTM D2241. Philadelphia, Pa.: American Society for Testing and Materials.

Atkins, A.G. and Y.W. Mai, 1988 *Elastic and Plastic Fracture*, Ellis Horwood, Chichester, UK.

AWWA (American Water Works Association). 1994. *Underground Installation of Polyvinylchloride (PVC) Pressure Pipe and Fittings for Water*, AWWA C605. Denver, Colo.: American Water Works Association.

AWWA (American Water Works Association). 2002. *PVC Pipe – Design and Installation*, AWWA M23. Denver, Colo.: American Water Works Association.

Aylett, G.J. 1990. Taking Rigid PVC into the 21st Century – a Breakthrough Approach. Paper No. 24 in *PVC 90 Conf., April 24–26, 1990, Brighton, UK*. The Plastics and Rubber Institute.

Bacaloglu, R., and M. Fisch. 1995. Degradation and Stabilisation of PVC: V – Reaction Mechanism of PVC Degradation. *Polym. Degrad. Stabil.*, 47:33–57.

Baker, P., and R.F. Grossman. 1996. Heat Stabilisers of PVC Compositions Containing a Mixed Metal Stabiliser and C1–C8 Dialkyltin Oxide. U.S. Patent 5,561,182.

Bancon–Montigny, C., G. Lespes, and M. Potin–Gautier. 2001. Optimisation of the Storage of Natural Freshwaters Before Organotin Speciation. *Water Research*, 35(1):224–232.

Barker, M.B., J. Bowman and M. Bevis. 1983. The performance and causes of failure of polyethylene pipes subjected to constant and fluctuating internal pressures. *Jour. Mater. Sci* 18:1095-1118

Batt, J.M. 2002. Tin Stabilisers: A History of Safe Use [Online]. Available: <http://www.atofinachemicals.com/additives/literature/tinstabilizers_AhistoryofSafeUse.pdf>. [cited May 14, 2004]

Batt, J.M. 2003. The World of Organotin Chemicals: Applications, Substitutes, and the Environment [Online]. Available: <http://www.ortepa.org/aboutorganotin.htm>. [cited May 14, 2004]

Beekman, G.F. 1996. Polyol Esters of Zinc Mercaptoacids as Heat Stabilisers for PVC Processing. U.S. Patent 5,536,767.

Benavides, R., M. Edge, N.S. Allen, and M. Shah. 1997. The Mode of Action of Metal Stearate Stabilisers in PVC: IV – The Application of Fluorescence Spectroscopy to Characterise Chromophoric Species. *Polym. Degrad. Stabil.*, 57:25.

Benham, P.P., and S.J. Hutchinson. 1970. *Plasts Polym.*, 259.

Benjamin, P. 1980. *Jour. Vinyl Technol.*, 2:254.

Berger, K. 1989. Pipe Extrusion – Alternatives to Lead Stabilisation are Already Available. *Plastverarbeiter*, 40(4):188–192.

Bilby, B.A., A.H. Cottrell and K.H. Swindon, 1963. *Proc. Roy. Soc. London*, **A272**, 304.

Billingham, N.C., and P.D. Calvert. 1980. In *Developments in Polymer Stabilisation*, Edited by G. Scott. London: Applied Science Publishers.

Bjorklund, I. 1992. In *Proc. of the Eighth International Conference on Plastics Pipes, Eindhoven, The Netherlands*. Plastics and Rubber Institute.

Bjorklund, I., and L.E. Janson. 1981. In *Proc. of the Int. Conf. on Underground Plastic Pipe, New Orleans, USA*. Reston, VA: American Society of Civil Engineers.

Balkose, D., H.I. Gokcel, and S.E. Goktepe. 2001. Synergism of Ca/Zn Soaps in Poly (Vinyl Chloride) Thermal Stability. *European Polymer Jour.*, 37:1191–1197.

Blom, G. 1989. Sannolikhetsteori och statistikteori med tillämpningar. Studentlitteratur, Lund, Sweden.

Boettner, E.A., G.L. Ball, Z. Hollingsworth, and R. Aquino. 1982. *Organic and Organotin Compounds Leached From PVC and CPVC Pipe*. Report EPA-600/S1-81-062.

Boraiko, C., M. Grasso, J. Batt, and G. Knowles. 2000. Evaluation of Employee Exposure to Butyltin Compounds Used as Tin Stabilisers at Polyvinyl Chloride Processing Facilities. Available: <http://www.plastics.ca/StaticContents/StaticPages/vcc/emp/vcc%20Study20on20%Emp%20Exposure%20to%Butyltin%20Compounds.pdf>.

Bradley, W.L., J.B. Slay and R.A. Self, 1998 In *Proc. of the Tenth International Conference on Plastics Pipes, Goteborg, Sweden*, pp. 867–874 IOM Communications.

Breant, P., and C.H. Stapfer. 1994. Prediction of Heat Stabiliser Consumption in PE, PP, PS and PVC by Fluorescence Reflectometry. In *Proc. of the ANTEC Conference,* pp. 2151–2155.

Broberg, K.B. 1968. *Int. Jour. Fract.*, 4:11.

Brogden, S. 2000. Ph.D. diss., Manchester Metropolitan University, UK.

Brogden, S., M.A. Shepherd, and G.P. Marshall. 1989. *Plastics and Rubber Processing and Applications*, 27:483.

Broutman, L.J., D.E. Duvall, and P.K. So. 1989. Failure Analysis of a PVC Water Pipe. In *Proc. of the Forty-Seventh Annual Technical Conference of the Society of Plastics Engineers, ANTEC '89*. Brookfield, CT: Society of Plastics Engineers.

BSI (British Standards Institute). 1973. *Code of Practice for Plastics Pipework (Thermoplastics Materials)*, CP 312. London, UK: British Standards Institute.

BSI (British Standards Institute). 1986. *Specification for Unplasticized Polyvinyl Chloride (PVC-U) Pressure Pipes for Cold Potable Water*, BS 3505. London, UK: British Standards Institute.

BSI (British Standards Institute). 1991. *Methods of Testing Plastics: Thermal Properties – Determination of the Thermal Stability of Polyvinyl Chloride by the Congo Red Method: Part 1 – Method 130A*, BS 2782. London, UK: British Standards Institute.

BSI (British Standards Institute). 1999. *Unplasticised Polyvinylchloride (PVC-U) Profiles for the Fabrication of Windows and Doors – Determination of the Resistance to Artificial Weathering*, BS EN 513. London, UK: British Standards Institute.

BSI (Bristish Standards Institute). 2000. *Suitability of Non-metallic Products for Use in Contact With Water Intended for Human Consumption With Regard to Their Effect on the Quality of the Water – Specification*, BS 6920-1. London, UK: British Standards Institute.

Bucknall, C.B. 2000. In *Polymer Blends: Volume 2 – Performance*. New York: John Wiley and Sons.

Bucknall, C.B., and P. Dumpleton. 1985. Factors Affecting Fatigue Crack Growth in HDPE. *Plast. Rubb. Proc and Appl.*, 5, 343–347.

Burgess, R.H. 1982. *Manufacturing and Processing of PVC*. Essex, UK: Applied Science Publishers.

Burley, J.W. 1997. New Organic Co-stabilisers for the Stabilisation of PVC. *Jour. Vinyl and Additive Technol.*, 3(3):205.

Burn L.S. (1991). The effect of installation damage on the lifetimes of UPVC pipes subjected to cyclic pressures, *Polymer International 26, 147-150.*

Burn, L.S. 1992. The Influence of Titanium Dioxide on the Durability of uPVC Pipe. *Polym. Degrad. Stabil.*, 36:155.

Burn, L.S. 1993. *Installation Damage: Effect on Lifetime of uPVC Pipes Subjected to Cyclic Pressure*. Research Report 68. Melbourne, Victoria: Urban Water Research Association of Australia.

Burn, L.S. 1994. Lifetime Prediction of uPVC Pipes – Experimental and Theoretical Comparisons. *Plast., Rubber Compos. Process. Appl.*, 21(2): 99–108

Burn, L.S. and Richards, T.J. 1995. Factors Affecting the Fracture Toughness Measurement of uPVC Pipe Samples. In *Proc. of the Ninth International Conference on Plastics Pipes, Edinburgh, UK*, pp. 223–232 IOM Communications

Burn, L.S., and B.L. Schafer, 1997. *The Environmental Impact of Lead Leaching From uPVC Sewerage, Waste and Vent Pipes*. Technical Report TR97/1. Victoria, Australia: CSIRO Building, Construction & Engineering.

Burn L.S., and A.P. Sullivan. 1993. New uPVC Water Reticulation Mains – The Effect of Commissioning Procedures on the Levels of Extracted Lead. *Jour. Water SRT – Aqua*, 42(3):135–145.

Butters, G., and G.C. Marks. 1978. Techniques for Predicting the Weathering Performance of Rigid PVC. *Jour. Macromol. Sci. – Chem., A*, 12(4):569–586.

Capocci, G., and J. Zappla, 1986. The Performance of Hindered Amine Light Stabilisers in Flexible PVC. In *Proc. of the Society of Plastics Engineers Forty-Third Annual Technical Conf.*, p. 1219.

Carr, D.S. 1978. Outdoor Weathering of PVC. *Elastomerics*, February:15–18.

Chapman, P.G., and L. Ågren. 1998. A New Technology for In-Line Manufacturing of Biaxially Oriented PVC Pipes. In *Proc. of Plastics Pipes X, September, 14–17 1998, Gothenburg, Sweden*. IOM Communications.

Christie, M.A., and B. Phelps. 1984. *Int. Jour. Fatigue*, 6:125.

Clutton, E. 2001. Essential Work of Fracture. In *Fracture Mechanics Testing Methods for Polymers, Adhesives and Composites.* Edited by D.R. Moore, A. Pavan, and J.G. Williams. ESIS Publication 28. Amsterdam: Elsevier.

Cooke, G.M. 2002. Effect of Organotins on Human Aromatase Activity in Vitro. *Toxicology Letters*, 126:121–130.

Cotterell and Mai. 1982 *Advances in Fracture Research (ICF-5) Vol. 4*. London: Pergamon Press.

Crema, J., and G.J. Aylett. 2002. In *Proc. of PVC 2002 Towards a Sustainable Future Conference, April, 2002, Brighton, UK*, pp. 410–415.

Crompton Vinyl Additives. 2001. Organic-Based Heat Stabilisers for Rigid PVC. *Plastic Additives and Compounding*, November:26–30.

Crowder, M.J., Kimber, A.C., Smith, R.L., and Sweeting, T.J. 1991. *Statistical Analysis of Reliability Data*, Chapman and Hall, London.

Cuilleret, M. 1998. Solid Sulphur Free Tin Stabilisers: New Alternatives to Stabilise Rigid PVC Applications. Paper No. 11 in *Proc. of Addcon World.*

Davis, P., L.S. Burn, and A.J. Whittle. 2001. Investigating Crack Growth and Plasticity in the C Ring Fracture Toughness Test. In *Proc. of Plastics Pipes XI, September 3–6, 2001, Munich, Germany*. IOM Communications.

Davis, P., B.E. Tiganis, and L.S. Burn. 2004. The Effect of Photo-Oxidation Degradation on Fracture in ABS Pipe Resins. *Polym. Degrad. Stabil.*, 84(2):233–242.

Defosse, M. 2002. Ca-Zn and Organics Aim to Take the Lead Out. *Modern Plastics*, July:40–41.

Deshmukh, R., and F. Lai. 2000. Quantitative Relationships Between the Parameters of Thermal Degradation of Polyvinylchloride and the Loss of Mechanical Properties. In *ANTEC Conf. Proc.*, 3421–3425.

Drewes, R., M. Kolb, K.J. Kuhn, H.J. Sander, and W. Wehner. 1996. Stabilised PVC. U.S. Patent 5,519,077.

Dukes, B.W. 1985. In *Proc. of the Sixth International Conference on Plastics Pipes, March, 1985, York, UK*. Plastics and Rubber Institute.

Ebdon, L., S.J. Hill, and C. Rivas. 1998. Organotin Compounds in Solid Waste: a Review of Their Properties and Determination Using High-Performance Liquid Chromatography. *Trends in Analytical Chem.*, 17(5):277–288.

Eckstein, D. 1988. Twenty Year Old PVC Pressure Pipe Excavation and Evaluation. In *1988 Annual Conference Proc.*, pp. 809–816. Denver, Colo.: American Water Works Association.

Elinck, J.P., J.C. Bauwens, and G. Holmes. 1971. *Int. Jour. Fract. Mech.*, 7:277.

Engel, L., H. Klingele, G.W. Ehrenstein and H. Schaper. 1978. *An Atlas of Polymer Damage*, Carl Hanser Verlag, London, UK.

Eskes, C., P. Honegger, T. Jones–Lepp, K. Varner, J.M. Matthieu, and F. Monnet–Tschudi. 1999. Neurotoxicity of Dibutyltin in Aggregating Brain Cell Cultures. *Toxicology in Vitro*, 13:555–560.

ESPA (2002) European Stabilizer Producers Association. Available: http://www.stabilizers.org [cited, September, 2002].

Europa. 1998. Drinking Water Directive 98/83/EC on the Quality of Water Intended for Human Consumption [Online]. Available: <http://www.waterunc.com/gb/indexgb.htm>.

Ewalds, H.L., and R.J.H. Wanhill. 1996. *Fracture Mechanics.* London, UK: Arnold.

Fedoseeva, G.T., L.D. Strelkova, E.O. Krats, V.P. Lebedev, and K.S. Minsker. 1981. *Int. Polym. Sci. and Tech.*, 8:T1.

Fent, K. 1996. Organotin Compounds in Municipal Wastewater and Sewage Sludge: Contamination, Fate in Treatment Process and Ecotoxicological Consequences. *Sci. Total Environ.*, 185:151–159.

Fitzpatrick, P., P. Mount, G. Smyth, and R.C. Stephenson. 1996. *Plast. Rubber Process. Applic.*, 25:359.

Forsyth, D.S., and B. Jay. 1997. Organotin Leachates in Drinking Water From Chlorinated Poly(Vinyl Chloride) (CPVC) Pipe. *App. Organometal. Chem.*, 11:551–558.

Franklin, G. 2002. The UK and European Acceptance Scheme: the Effects of Materials on Water Quality. *Water Conditioning and Purification* [Online], September:42–46. Available: <http://www.wcp.net/pdf/0902franklin.pdf>. [cited July 6, 2004]

Gachter, R and H. Muller. 1985, Plastics Additives Handbook, Hanser, Munich, Germany.

Gardette, J.L., S. Gaumet, and J.L. Philipart. 1993. Influence of the Experimental Conditions on the Photooxidation of Poly(Vinyl Chloride). *Jour. Appl. Polym. Sci.*, 48:1885.

Gardette, J.L., and J. Lemaire. 1991. Photothermal and Thermal Oxidation of Rigid Plasticised and Pigmented Poly(Vinyl Chloride). *Polym. Degrad. Stabil.*, 34:155–167.

Gesenhues, U. 2000. Influence of Titanium Dioxide Pigments on the Photo-Degradation of Poly(Vinyl Chloride). *Polym. Degrad. Stabil.*, 68:185–196

Gesenhues, U., and J. Hocken. 2000. Titanium Dioxide Pigment Structure and Kinetics of PVC Weathering. *Jour. Vinyl Additive Tech.,* 6(2):80–87.

Gons, J., 1995. 35 Years of PVC Water Distribution Pipes in the Netherlands. *Uni-Bell PVC Pipe News*, Spring:9–11.

Gotham, K.V. 1969. *Plast. Plasts. Polym.*, 309.

Gotham, K.W., and M.J. Hitch. 1978. *Brit. Polym. Jour.* 10:47.

Greenshields, C.J., P. Davis, and P.S. Leevers. 1997. *Plast. Rubber Process. Applic.*, 26:404.

Gumargalieva, K.Z., V.B. Ivanov, G.E. Zikov, T. Moiseev, V. Ju, and V. Pokholok. 1996. Problems of Ageing and Stabilisation of Poly(Vinyl Chloride). *Polym. Degrad. Stabil.*, 52:73–79

Hattori, T., K. Tanaka, and M. Matsuo, 1972. *Polym. Eng. Sci.*, 12:199.

Hay, J.N. 1970. *Jour. Polym. Sci., A1*, 9:1823.

Hertzberg, R.W., and J.A. Manson. 1980. *Fatigue of Engineering Plastics*. New York: Academic Press.

Hoch, M. 2001. Review: Organotin Compounds in the Environment – an Overview. *Appl. Geochem.*, 16:719–743.

Holloway, L.R., and A.J. Naaktgeboren. 1992. In *Proc. of the Eighth International Conference on Plastics Pipes*. Eindhoven, The Netherlands: Plastics and Rubber Institute.

Holmes, R. 2004. Personal communication. City of Ottawa, Infrastructure Services Branch.

Hopfmann, Th. 2002. Organic Based Stabiliser System – Commercial Introduction. In *Proc. of the PVC 2002 Towards a Sustainable Future Conference, April, Brighton, UK*, pp. 515–528.

Hruza, A. 2002. Chemson Pacific, Personal Communication.

HULS. 1985. *Requirements of Windows and Window Materials: Test Methods and Results – Economy*. 2nd ed., pp. 6–18. Germany: Chemische Werke HULS AG.

ISO (International Organization for Standardization). 1977. *Unplasticised Polyvinyl Chloride (PVC) Pipes for Potable Water Supply: Extraction of Lead and Tin – Test Method*. ISO 3114. Geneva, Switzerland: International Organization for Standardization.

ISO (International Organization for Standardization). 1990. *Plastics: Determination of Thermal Stability of Poly(Vinyl Chloride), Related Chlorine-Containing Homopolymers and Copolymers and Their Compounds – Discoloration Method*. ISO 305. Geneva, Switzerland: International Organization for Standardization.

ISO (International Organization for Standardization). 1993. *Plastics: Determination of the Tendency of Compounds and Products Based on Vinyl Chloride Homopolymers and Copolymers to Evolve Hydrogen Chloride and Any Other Acidic Products at Elevated Temperatures: Part 3 – Conductometric Method*. ISO 182-3. Geneva, Switzerland: International Organization for Standardization.

ISO (International Organization for Standardization). 1994. *Thermoplastics Pipes – Determination of Resistance to External Blows – Staircase Method*. ISO 11173. Geneva, Switzerland: International Organization for Standardization.

ISO (International Organization for Standardization). 1995. *Unplasticized Poly(Vinyl) Chloride (PVC-U) Pipes – Dichloromethane Resistance at Specified Temperature (DCMT) – Test Method*. ISO 9852. Geneva, Switzerland: International Organization for Standardization.

ISO (International Organization for Standardization). 1999. *Determination of the Fracture Toughness Properties of Un-plasticized Poly(Vinyl Chloride) (PVC-U) Pipes*. ISO/DIS 11673.2. Geneva, Switzerland: International Organization for Standardization.

ISO (International Organization for Standardization). 2001. *Plastic Piping Systems for the Transport of Water Intended for Human Consumption: Migration Assessment – Determination of Migration Values for Plastic Pipes and Fittings and Their Joints*. ISO 8795. Geneva, Switzerland: International Organization for Standardization.

ISO (International Organization for Standardization). 2003. *Plastics Pipes and Ducting Systems – Determination of the Long-Term Hydrostatic Strength of thermoplastics Materials in Pipe Form by Extrapolation*. ISO 9080. Geneva, Switzerland: International Organization for Standardization.

Jarrett, R., O. Hussain, A. Veevers, and J. Van der Touw. 2001. In *Proc. of the International Conference of Maintenance Societies, Melbourne, Australia*.

Jeffrey, J.D., A.P. Moser and S.L. Folkman. 2004. Long Term Cyclic Testing of PVC Pipe – Final Report. Completed for Uni-Bell PVC Pipe Association. Accessed from www.uni-bell.org on 11th November, 2004

Jones–Lepp, T.L., K.E. Varner, and B.A. Hilton. 2001. *Appl. Organometal. Chem.*, 15:933–938.

Joseph, S.H. 1984. Fatigue Failure and Service Lifetime in uPVC Pressure Pipes. *Plast. Rubber Process. Applic.*, 4:325–330

Joseph, S.H., and P.S. Leevers. 1985. *Jour. Mater. Sci.*, 20:237.

Jouan, X., and J.L. Gardette. 1987. *Polymn. Comm.*, 28:329.

Kaninen, M.F., P.E. O'Donoghue, C.H. Popelar, and V.H. Kenner. 1990. A Visco-Elastic Fracture Mechanics Assessment of Slow Crack Growth in Polyethylene Gas Distribution Pipe Materials. *Eng. Fract. Mech.*, 36(6):903–918

Kim, H.S., R.W. Truss, Y.W. Mai and B. Cotterell. 1988. Fatigue Crack Propagation in Unplasticised Polyvinyl Chloride (uPVC) Part 1: Effect of Mean Stress, *Polymer*, 29:268

Kirby, P.C. 1981. In *Proc. of International Conference on Underground Plastic Pipe*. New Orleans: American Society of Civil Engineers.

Kuraoka, S., B. Rajani and C. Zhan, 1996. Pipe-Soil Interaction Analysis of Field Tests of Buried PVC Pipe. Jour. Infrastruct. Sys. 2(4):162-170

Kwon, J.A., and R.W. Truss. 2002. The Work of Fracture in Uniaxial and Biaxial Oriented Unplasticised Polyvinylchloride Pipes. *Eng. Fract. Mech.*, 69:605–616

Lau, S.W., and R.W. Truss. 2002. The Essential Work of Fracture Through the Wall of a uPVC Pipe. *Jour. Mater. Sci.*, 37:1115–1119

Lawrence, C., S. Teo, and R. Potter. 1998. In *Proc. of Plastics Pipes X Conference, September 14–17, 1998, Gothenburg, Sweden*. IOM Communications.

Levita, G., L. Parisi, A. Marchetti, and L. Bartolommei. 1996. Effect of Thickness on the Specific Essential Work of Fracture of Rigid PVC. *Polym. Eng. Sci.*, 36(20):2534–2541.

Lewis, B. 1993. Mathematical Models for the Durability of Plastics. In *Proc. of the Society of Plastics Engineers, Vinyl Division Vinyl 20, Regional Technical Conference SPE*, pp. 131–142. Brookfield, CT: Society of Plastics Engineers.

Linder, R.A., and K. Worschech. 1986. Lubricants for PVC. In *Encyclopedia of PVC*. 2nd ed., pp. 263–390. Edited by L.I. Nass, and C.A. Heiberger. New York: Marcel Dekker.

Lu, J.P., P. Davis, and L.S. Burn. 2003. Lifetime Predictions for ABS Pipes Subjected to Combined Pressure and Deflection Loading. *Polym. Eng. Sci.*, 43:444.

MacKellar, S. and D. Pearson. 2003. Nationally Agreed Failure Data and Analysis Methodology for Water Mains Volume 1: Overview and Findings. Report Ref. No. 03/RG/05/7, UK Water Industry Research, London, UK.

Mai, Y-W. and Cotterell, B. (1986). On the Essential Work of Ductile Fracture in Polymers. *Int. Jour. Fract.*, 32:105–25

Marshall, G.P. 1982. Design for Toughness in Polymers: 1 – Fracture Mechanics. *Plast., Rubber Compos. Process. Appl.*, 2(2):169–182.

Marshall, G.P. 1997. Plastics Pipeline Systems Evaluation of Long-term Performance – The Behaviour of Buried Pipes. Ref. No. 97/WM/02/5, UK Water Industry Research, London, UK.

Marshall, G.P., S. Brogden, and M.A. Shepherd. 1998. Evaluation of the Surge and Fatigue Resistance of PVC and PE Pipeline Materials for Use in the U.K. Water Industry. In *Proc. Plastics Pipes X Conference, September 14–17, 1998, Gothenburg, Sweden*. IOM Communications.

Marshall, G.P., D. Pearson, and S. MacKellar. 1996. Specification of Performance for Modified Poly(Vinyl Chloride) and Polyethylene Pipes. *Plast., Rubber Compos. Process. Appl.*, 25(6):276–286

Marshall, G.P., and J.G. Williams, 1974. In *Proc. of the Third International Plastic Pipes Symposium, Southampton, UK.* London: British Plastics Federation.

Martin, K.G. 1977. *Solar Weathering Indices for Australian Sites*. Research Technical Paper 18. Melbourne, Australia: CSIRO Building, Construction & Engineering.

Maumud, F., and I. Hussain. The Suitability of Saudi Arabian, ASTM, BS and DIN Standards for uPVC Pipes Used Under Local Environmental Conditions. *Full reference to be supplied.*

Melchers, R.E. 1999 *Structural Reliability Analysis and Prediction – 2nd Edition*, John Wiley and Sons, Brisbane.

Menges, G., and N. Berndtsen. 1976. *German Plastics Kunstoffe*, 66:9.

Menges, G., and N. Berndtsen. 1977. *Pure Appl. Chem.*, 49:597.

Menzel, G. 1979. Weathering Tests on PVC Modified With Ethylene-Vinyl-Acetate Copolymers (EVAC). *Kunststoffe*, 69(8):480–482.

Mesch, K.A., and T.G. Kugele. 1992. Use of Organotin Stabilisers – Risk Assessment Analysis. In *Proc. of ANTEC 1992 Conf.* pp. 1942–1945.

Michel, A., T.V. Hoang, B. Perrin, and M.F. Lauro. 1981. Synergistic Mechanisms of β-Diketone Derivatives and Zinc-Calcium Soaps in PVC Stabilisation. *Polym. Degrad. Stabil.*, 3:107–119.

Mills, N.J., and N. Walker. 1976. *Polymer*, 17:335.

Minsker K.S., S.V. Kolesov, and G.E. Zaikov. 1988. *Degradation and Stabilisation of Vinyl Chloride Based Polymers*. Oxford, UK: Pergamon Press.

Morel, P., M. Cuilleret, and S. Girois. 1999. New Solutions for Stabilisation of Weatherable Rigid Poly(Vinyl Chloride). *Plast., Rubber and Composites*, 28(4):175–179.

Mur, G.M. 1995. The Role of Beta Diketones in Ca/Zn Stabilising Systems for Weatherable and Rigid PVC. In *Proc. of ANTEC 1995 Conf.* pp. 3599–3607.

Naqvi, M.K. 1985. Structure and Stability of Poly(Vinyl Chloride). *JMS-REV. Macromol. Chem. Phys.*, 25(1):119–155.

Navio, J.A., F.J. Marchena, C. Cerrilos, and F. Pablos. 1993. UV Photolytic Degradation of Butyltin Chlorides in Water. *Jour. Photochem. Photobiol. A: Chem.*, 71(1):97–102.

NHMRC (National Health and Medical Research Council and Agriculture and Resource Management Council of Australia and New Zealand). 1996. *Australian Drinking Water Guidelines – Summary*. Canberra: Commonwealth of Australia.

Nielsen, J.B., and J. Strand. 2002. Butyltin Compounds in Human Liver. *Environ. Research Section A*, 88:129–133.

ORTEPA (Organotin Environmental Programme Association). 1997. *Human Health Aspects* [Online]. Available: http://www.ortepa.org/stabilizers/pages/humanhealthaspects.htm>. [cited July 6, 2004]

Osawa, Z., and M. Aiba. 1982. Initiation Mechanism of the Photo-Degradation of Poly(Vinyl Chloride). *Polym. Photochem.*, 2:397–407.

Packham, R.F. 1971a, The Leaching of Toxic Stabilizers from PVC water pipe: Part I - A critical study of laboratory test Procedures., Water Treatment Exam., 20, pp. 108-124.

Packham. R.F. 1971b. The Leaching of Toxic Stabilizers from PVC water pipe: Part II - A survey of lead levels in uPVC distribution systems, Water Treatment Exam., 20, pp. 144-151

Parey, J., and G. Menges. 1981. Jour. Vinyl Technol., 3:152.

Paris, P.C. and G.C. Sih, 1963 *Stress Analysis of Cracks – Fracture Toughness Testing and its Applications*, American Society for Testing and Materials, Philadelphia, STP 381, p. 30.

Parks, D. 1979 *Int. Jour. Fracture*, 10:487

Parsons, M., E.V. Stepanov, A. Hiltner, and E. Baer. 2000. Correlation of Fatigue and Creep Slow Crack Growth in a Medium Density Polyethylene Pipe Material. *Jour. Mat. Sci.*, 35:2674.

Pfaender, R., W. Wehner, Å. Ryningen, and J.U. Stoffelsma. 1998. Organic-Based PVC Stabilizer. *Kunstoffe Plast Europe*, 88:704.

Pielichowski, K. and I. Hamerton. 1998. TGA/FTIR Studies on the Thermal Stability of PVC Blends With a Novel Colourant and Stabiliser: 3-(2,4-Dichlorophenylazo)-9-(2,3-Epoxypropanbe) Carbazole. *Polymer*, 39(1):241.

Potente. H., and S.M. Schultheis, 1987. *Kunstoffe German Plastics,* 77:19.

Prinet–Anton, C., J. Dubois, G. Mur, M. Gay, L. Audouin, and J. Verdu. 1998. Photoageing of Rigid PVC: II – Degradation Thickness Profiles. *Polym. Degrad. Stabil.*, 60:275–281.

Puterman, M. 1989. Natural and Accelerated Weathering of PVC and Polypropylene Waste-Water Pipes. *Materials and Structures*, 22:170.

Quevauviller, P. (1991); Leaching of organitin compounds from poly (vinyl chloride) (PVC material), Appl.OrganometallicChem 5, 125-129.

Qureshi, F.S., S.H. Hamid, and A.G. Maadhah. 1989. Weather Induced Degradation of Plastic Pipes. *Polym.-Plast. Technol. Eng.*, 28(7&8):663–670.

Rabek, J.F., B. Ranby, and T.A. Skowronski. 1985. Photothermal Dehydrochlorination of Poly(Vinyl Chloride). *Macromolecules*, 18:1810–1818.

Radon, J.C. 1977. *Jour. Macromol. Sci. – Phys. B,* 14(4):511.

Ragab, A.R., and H. Alawi. 2001. Weathering Effects on the Fracture Properties of Polyvinyl Chloride Pipes. *Jour. Testing & Evaluation*, 29(1):44–49

Rahman, S. 2002. PVC Pressure Pipes: Past, Present and Future. In *Proc. CIGMAT 2002*. Available: http://cigmat.cive.uh.edu/content/conf_exhib/02_present/10.html. Center for Innovative Grouting Materials & Technology.

Rajani, B., J. Makar, S. McDonald, C. Zhan, S. Kuraoka, C.K. Jen, and M. Viens. 2000. *Investigation of Grey Cast Iron Water Mains to Develop a Methodology for Estimating Service Life*. AwwaRF Project No. 280. Denver, Colo.: AwwaRF.

Rajani, B., S. McDonald, and G. Felio. 1995. *Water Mains Break Data on Different Pipe Materials for 1992 and 1993*. Ottawa, Ontario: National Research Council of Canada.

Rajani, B. and C. Zhan. 1996. On the Estimation of Frost Loads. *Can. Geotech. Jour*. 33:629-641

Real, L.P., and J.–L. Gardette. 2001a. Ageing and Characterisation of PVC-Based Compounds Utilised for Exterior Applications in the Building Construction Field: 1 – Thermal Ageing. *Polym. Testing*, 20:779–787.

Real, L.P., and J.–L. Gardette. 2001b. Ageing and Characterisation of PVC-Based Compounds Utilised for Exterior Applications in the Building Construction Field: 2 – Artificial Accelerated Ageing With Xenon Light. *Polym. Testing*, 20:789–794.

Richardson, K.A. 2002. In *Proc. 36th Hawaii Conf. On System Sciences*, Hawaii, USA.

Rooke D.P. and Cartwright D.J. 1976. *Compendium of Stress Intensity Factors*. London, UK: HMSO.

Research Institute King Fahid University of Petroleum and Minerals. 1985. *Testing PVC Pipe*. Dhahran for Royal Commission of Jubai and Yanbu Saudi Arabia.

Rosenthal, M. 2002. The Voluntary Commitment – Stabiliser Changes. In *Proc. of PVC 2002 Towards a Sustainable Future Conf., April, 2002, Brighton, UK*. pp. 53–63.

Sadiki, A.–I., and D.T. Williams. 1996. Speciation of Organotin and Organolead Compounds in Drinking Water by Gas Chromatography–Atomic Emission Spectrometry. *Chemosphere*, 32(10):1983–1992.

Sadiki, A.–I., and D.T. Williams, 1999. A Study on Organotin Levels in Canadian Drinking Water Distributed Through PVC Pipes. *Chemosphere*, 38(7):1541–1548.

Sadiki, A.–I., D.T. Williams, R. Carrier, and B. Thomas. 1996. Pilot Study on the Contamination of Drinking Water by Organotin Compounds from PVC Materials. *Chemosphere*, 32(12):2389–2398.

Saleemi, A.S. and J.A. Nairn. 1990. The Plane-Strain Essential Work of Fracture as a Measure of the Fracture Toughness of Ductile Polymers. *Polym. Eng. Sci.* 30(4):211-218

Schiller, M. and W. Fischer, 2004. Stabilizers for PVC pipe systems - Quo Vadis? [online]. Available: http://www.pvc4pipes.org/documents/files/PXII/sess3b/Schiller.pdf

Scrima, R., and G. Vessa. 1995. Stabilisation of Polymeric Organic Materials by Using Synergistic Mixtures Comprising Sterically Hindered Cyclic Amines and Derivatives of 3-Pyrazolidinone or 1,2,4-Triazolidine-3,5-Dione. U.S. Patent 5,457,143I.

Seargant, D. 1999. Reviewing the City of Edmonton Experience. *Uni-Bell PVC Pipe News*, Spring:1–3.

Seldén, R. 1987. The Effect of Heat Aging on Fracture Toughness in uPVC Pressure Pipe. *Polym. Testing*, 11:431–444.

Shih, C.F. 1985. J-Dominance Under Plane Strain Fully Plastic Conditions. *Int. Jour. Fracture*, 29:73–84.

Stahmer, M.W., and A.J. Whittle. 2001. In *Proc. of Plastics Pipes XI Conference, September 3–6, 2001, Munich, Germany*, IOM Communications.

Standards Australia. 1979. *Installation of PVC Pipe Systems*, AS 2032 (Committee Draft). Sydney, Australia: Standards Australia.

Standards Australia. 1984. *Plastics Pipes and Fittings For Gas Reticulation – Unplasticised PVC (PVC-U): Part 1 – Pipes; and Part 2 – Fittings*, AS 1464.1–2. Sydney, Australia: Standards Australia.

Standards Australia. 1990. *Methods for Impact Tests on Plastics: Part 2 – Charpy Impact Resistance*, AS 1146.2. Sydney, Australia: Standards Australia.

Standards Australia. 2003. *Plastics – Determination of Impact Properties: Part 2 – Instrumented Charpy Impact Tests*, DR 0338 CP. Sydney, Australia: Standards Australia.

Standards Australia. 2003. *Oriented PVC (PVC-O) Pipes for Pressure Applications*, AS 4441(Int). Sydney, Australia: Standards Australia.

Standards Australia/Standards New Zealand, 1998, *Buried flexible pipelines Part 1: Structural Design*, AS 2566.1.,Sydney Australia, Standards Australia

Standards Australia/Standards New Zealand. 1999. *PVC Pipes and Fittings for Pressure Applications*, AS/NZS 1477. Sydney, Australia: Standards Australia.

Standards Australia/Standards New Zealand. 2000a. *Products for Use in Contact With Drinking Water*, AS/NZS 4020. Sydney, Australia: Standards Australia.

Standards Australia/Standards New Zealand. 2000b. *Modified PVC (PVC-M) Pipes for Pressure Applications*, AS/NZS 4765 (Int). Sydney, Australia: Standards Australia.
Standards Australia/Standards New Zealand. 2003. *Methods of Test for Plastics Pipe and Fittings: Method 19 – C-Ring Test for Fracture Toughness of PVC Pipes*, AS/NZS 1462.19(Int). Sydney, Australia: Standards Australia.
Stapel, J.J. 1977. *Pipes & Pipelines Int.*, February:11.
Starnes, W.H. 1981. *Dev. Polym. Deg.*, 3:135.
Swasey, C. 1980. An Updated Guide to UV Stabilisation, *Plast. Eng.*, August:33.
Teh, J.W., A.A. Cooper, A. Rudin, and J.L.H. Baptiste. 1989. *Jour. Vinyl Technol.*, 11(1):33.
Teh, J.W., A.A. Cooper, A. Rudin, and J.L.H. Baptiste. 1989. *Makromol. Chem., Macromol. Symp.*, 29:123.
Thoft-Christensen, P. and M.J. Baker. 1982. *Structural Reliability Theory and its Applications*, Springer-Verlag, Berlin.
Thomas, N.L. 1993. Calcium/Zinc Stabilisers for PVC Pressure Pipe: Review. *Plast. Rubber Compos. Process. Applic.*, 19:263–271.
Tiganis, B.E., L.S. Burn, P. Davis, and A.J. Hill. 2002. Thermal Degradation of Acrylonitrile–Butadiene–Styrene (ABS) Blends. *Polym. Degrad. Stabil.*, 76:425.
Truss, R.W. 1988. Lifetime Predictions for uPVC Pipes Subjected to Combined Mean and Oscillating Pressures. *Plast. Rubber Process. Applic.*, 10:1–9.
Uni-Bell. 1981. *UNI-TR-5: The Effects of Ultraviolet Aging on PVC Pipe*. Dallas, Texas: Uni-Bell PVC Pipe Association.
Uni-Bell. 2001. *Handbook of PVC Pipe, Design and Construction*. Dallas, Texas: Uni-Bell PVC Pipe Association.
Unibell, 2002, What is the life-expectancy of PVC pipe [online] Available: http://www.uni-bell.org/faq.html#q2,[cited, 16 December, 2002].
USEPA (U.S. Environmental Protection Agency). 2002. *National Primary Drinking Water Standards* [Online]. Available: <http://www.epa.gov/safewater>. [cited May 17, 2004]
Verdu, J. 1978. *Jour. Macromol. Sci. Chem., Part A*, 12:551.
Veronelli, M., M. Mauro, and S. Bresadola. 1999. Influence of Thermal Dehydrochlorination on the Photo-Oxidation Kinetics of PVC Samples. *Polym. Degrad. Stabil.*, 66:349–357.
Vinson, H.W. 1981. In *Proc. of the International Conference on Underground Plastic Pipe*. New Orleans: American Society of Civil Engineers.
Water UK. 2000. *Water Industry Specification WIS 4-37-02*. March, Issue 1. London: Water UK.
Weibull, W. 1948 En statistisk teori för utmattningshållfasthetslära. Aktiebolaget Bofors.
Whalen, M.W., and G. Longanathan. 2001. Butyltin Exposure Causes a Rapid Decrease in Cyclic AMP Levels in Human Lymphocytes. *Toxicol. & Appl. Pharmacol.*, 171:141–148.
Whittle, A.J. 2002. The Optimization of Rubber-Modified Poly(Vinyl Chloride) Pipes for Pressure Applications. Ph.D. diss., University of New South Wales, Sydney, Australia.
Whittle, A.J., and R.P. Burford. 1998. Optimizing Performance of Modified Poly(Vinyl Chloride) for Pressure Pipe Applications. *Plast. Rubber Compos. Process. Applic.* 27(9):395–399.
Whittle, A.J., R.P. Burford, and M. Hoffman. 2000. Improvements to C-Ring Fracture Toughness Test for Poly(Vinyl Chloride) Pipes. *Plast. Rubber Compos. Process. Applic.* 29(8):417–419.

WHO (World Health Organization). 1993. *Guidelines for Drinking-Water Quality: Vol. 1 – Recommendations*, 2nd ed., pp. 49–50. Geneva, Switzerland: World Health Organization. http://www.who.int/water_sanitation_health/GDWQ/Chemicals/leadsum.htm.

WHO (World Health Organization). 1996. *Guidelines for Drinking-Water Quality: Vol. 2 – Health Criteria and Other Supporting Information.* Geneva, Switzerland: World Health Organization. http://www.who.int/water_sanitation_health/GDWQ/Summary_Tables/Tab2d.htm.

Williams, J.G. 1978. Applications of Linear Fracture Mechanics. *Adv. Polym. Sci.*, 27:67–120.

Williams, J.G. 1984. *Fracture Mechanics of Polymers*, Ellis Horwood, Chichester, UK.

Yassin, A.A., M.W. Sabaa, and N.A. Mohamed. 1985. Polymerisation Products of p-Benzoquinone as Thermal Stabilisers for Rigid PVC: Part II – Evaluation of the Stabilising Efficiency. *Polym. Degrad. Stabil.*, 13:225–247.

ABBREVIATIONS

a	crack size
ABS	acrylonitrile–butadiene–styrene
AS	Australian Standards
ASTM	American Society for Testing and Materials
b	wall thickness (mm)
Ba	barium
BCS	Bilby, Cottrell, and Swinden
BSI	British Standards Institute
°C	degrees Celsius
C	carbon
Ca	calcium
C=C	carbon unsaturated carbon bond (diene)
CCH_3	carbon with methyl group substitution
C-Cl	carbon–chlorine bond
Cd	cadmium
CEN	European Commission Norm
Cl	chlorine
COD	crack opening displacement
CO	carbonyl group (oxidation byproduct)
CPE	chlorinated polyethylene
CPVC	chlorinated poly (vinyl) chloride
DENT	double edge notch tensile
DIN	Deutsche Institut fur Normung
DIT	dehydrochlorination induction time
DSC	differential scanning calorimetry
δ^*	critical crack opening displacement
Δ	delta or change in
E	Young's modulus
EDS	energy dispersive X-ray spectroscopy
EEC	European Economic Community
EN	European Standard
EPA	Environmental Protection Agency
ESPA	European Stabilisers Producers Association
EU	European Union
EVAs	ethylene vinyl actetates
EWF	essential work of fracture
F	safety factor
FEA	finite element analysis

FOSM	first-order second-moment
FTIR	Fourier transform infra-red
HCl	hydrochloric acid
HDB	hydrostatic design basis
ISO	International Standards Organization
J	crack resistance
K_I	applied stress intensity factor (SIF)
K_{IC}	fracture toughness
K_{ID}	applied SIF from diametrical deflection
K_{IP}	applied SIF from internal pressure
K_{IR}	applied SIF from residual stress
K_{ITOT}	total applied SIF
LAB	yellow–red/green–blue indices of the color scheme
LCL	lower confidence limit
LEFM	linear elastic fracture mechanics
LTHS	long-term hydrostatic design stress
MPa	megapascals
NHMRC	National Health and Medical Research Council
OBS	organic-based stabilizer
OIT	oxygen induction time
Pb	lead
pphr	parts per hundred resin
PrEN	Provisional European Standard
PVC	polyvinyl chloride
PVC-M	modified polyvinyl chloride
PVC-O	oriented polyvinyl chloride
PVC-U	unplasticized polyvinyl chloride
P24D	pyrrolidine-2,4-dione
RH	relative humidity
S	
SBR	styrene–butadiene–rubber
SEM	scanning electron microscope/microscopy
SENB	single edge notch bend
SIF	stress intensity factor

Sn	tin
S:V	surface to volume ratio
σ	remotely applied stress
σ_Y	yield stress
TDS	total dissolved solids
T_g	glass transition temperature
TGA	thermogravimetric analysis
t_i	incubation time for crack initiation
T_iO_2	titanium dioxide
TOC	total organic carbon
T_s	time of stability (minutes)
UK	United Kingdom
UV	ultraviolet
UVA	ultraviolet absorber
WHO	World Health Organization
Y	fracture mechanics geometry factor
YVW	Yarra Valley Water
Zn	zinc
$ZnCl_2$	zinc chloride